CATALYSIS FOR ENERGY

Fundamental Science and Long-Term Impacts of the U.S. Department
of Energy Basic Energy Sciences Catalysis Science Program

Committee on the Review of the Basic Energy Sciences Catalysis
Science Program

Board on Chemical Sciences and Technology
Division on Earth and Life Studies

NATIONAL RESEARCH COUNCIL
OF THE NATIONAL ACADEMIES

THE NATIONAL ACADEMIES PRESS
Washington, D.C.
www.nap.edu

NATIONAL ACADEMIES PRESS
500 Fifth Street, NW, Washington, DC 20001

NOTICE: The project that is the subject of this report was approved by the Governing Board of the National Research Council, whose members are drawn from the councils of the National Academy of Sciences, the National Academy of Engineering, and the Institute of Medicine. The members of the committee responsible for the report were chosen for their special competences and with regard for appropriate balance.

This study was supported by the U.S. Department of Energy under Grant DE-AM01-04PI4503 (Task Order DE-AT01-07ER15924.00). Any opinions, findings, conclusions, or recommendations expressed in this publication are those of the authors and do not necessarily reflect the views of the organizations or agencies that provided support for the project.

International Standard Book Number-13: 978-0-309-12856-8
International Standard Book Number-10: 0-309-12856-0

Additional copies of this report are available from the National Academies Press, 500 Fifth Street, NW, Lockbox 285, Washington, DC 20055; (800) 624-6242 or (202) 334-3313 (in the Washington metropolitan area); Internet, http://www.nap.edu.

Cover: Design by Van Ngyuen; TOP IMAGE: Getty Images—petroleum-refining for energy and chemicals relies heavily on catalysis. BOTTOM IMAGE: Scanning electron micrograph of manganese-oxide molecular sieves—manganese oxides are used extensively in chemical processes for ion-exchange, separation, catalysis, and energy storage in secondary batteries. Courtesy of Steven L. Suib, University of Connecticut

Copyright 2009 by the National Academy of Sciences. All rights reserved.

Printed in the United States of America

THE NATIONAL ACADEMIES
Advisers to the Nation on Science, Engineering, and Medicine

The **National Academy of Sciences** is a private, nonprofit, self-perpetuating society of distinguished scholars engaged in scientific and engineering research, dedicated to the furtherance of science and technology and to their use for the general welfare. Upon the authority of the charter granted to it by the Congress in 1863, the Academy has a mandate that requires it to advise the federal government on scientific and technical matters. Dr. Ralph J. Cicerone is president of the National Academy of Sciences.

The **National Academy of Engineering** was established in 1964, under the charter of the National Academy of Sciences, as a parallel organization of outstanding engineers. It is autonomous in its administration and in the selection of its members, sharing with the National Academy of Sciences the responsibility for advising the federal government. The National Academy of Engineering also sponsors engineering programs aimed at meeting national needs, encourages education and research, and recognizes the superior achievements of engineers. Dr. Charles M. Vest is president of the National Academy of Engineering.

The **Institute of Medicine** was established in 1970 by the National Academy of Sciences to secure the services of eminent members of appropriate professions in the examination of policy matters pertaining to the health of the public. The Institute acts under the responsibility given to the National Academy of Sciences by its congressional charter to be an adviser to the federal government and, upon its own initiative, to identify issues of medical care, research, and education. Dr. Harvey V. Fineberg is president of the Institute of Medicine.

The **National Research Council** was organized by the National Academy of Sciences in 1916 to associate the broad community of science and technology with the Academy's purposes of furthering knowledge and advising the federal government. Functioning in accordance with general policies determined by the Academy, the Council has become the principal operating agency of both the National Academy of Sciences and the National Academy of Engineering in providing services to the government, the public, and the scientific and engineering communities. The Council is administered jointly by both Academies and the Institute of Medicine. Dr. Ralph J. Cicerone and Dr. Charles M. Vest are chair and vice chair, respectively, of the National Research Council.

www.national-academies.org

COMMITTEE ON THE REVIEW OF THE BASIC ENERGY SCIENCES CATALYSIS SCIENCE PROGRAM

NANCY B. JACKSON *(Co-chair)*, Sandia National Laboratory, Albuquerque, NM
JENS K. NØRSKOV *(Co-chair)*, Technical University of Denmark, Lyngby, Denmark
MARK A. BARTEAU, University of Delaware, Newark
MARK J. CARDILLO, Camille and Henry Dreyfus Foundation, New York
MARCETTA Y. DARENSBOURG, Texas A&M University, College Station
ANNE M. GAFFNEY, Lummus Technology, Bloomfield, NJ
VERNON C. GIBSON, Imperial College, London, England
SOSSINA M. HAILE, California Institute of Technology, Pasadena
MASATAKE HARUTA, Tokyo Metropolitan University, Japan
NENAD M. MARKOVIC, Argonne National Laboratory, IL
THOMAS A. MOORE, Arizona State University, Tempe
BRENDAN D. MURRAY, Shell Global Solutions, Houston, TX
JAMES C. STEVENS, Dow Chemical Company, Freeport, TX
BARRY M. TROST, Stanford University, Palo Alto, CA

Staff

DOROTHY ZOLANDZ, Director, Board on Chemical Sciences and Technology
ANDREW C. CROWTHER, Postdoctoral Research Associate
NORMAN GROSSBLATT, Senior Editor
TINA M. MASCIANGIOLI, Senior Program Officer
KELA L. MASTERS, Program Associate (through October 2008)
MICHAEL MOLONEY, Senior Program Officer, National Materials Advisory Board
ERICKA M. MCGOWAN, Associate Program Officer
JESSICA L. PULLEN, Administrative Assistant
SHEENA F. SIDDIQUI, Research Assistant
LYNELLE C. VIDALE, Program Assistant

BOARD ON CHEMICAL SCIENCES AND TECHNOLOGY

F. FLEMING CRIM *(Co-chair)*, University of Wisconsin, Madison
GARY S. CALABRESE *(Co-chair)*, Corning, Inc., NY
BENJAMIN ANDERSON, Eli Lilly K.K., Kobe, Japan
PABLO G. DEBENEDETTI, Princeton University, NJ
RYAN R. DIRKX, Arkema, Inc., King of Prussia, PA
MARY GALVIN-DONOGHUE, Air Products and Chemicals, Inc., Allentown, PA
PAULA T. HAMMOND, Massachusetts Institute of Technology, Cambridge
CAROL J. HENRY, Advisor and Consultant in Public Health and Environment, Bethesda, MD
RIGOBERTO HERNANDEZ, Georgia Institute of Technology, Atlanta
CHARLES E. KOLB, Aerodyne Research, Inc, Billerica, MA
MARTHA A. KREBS, California Energy Commission, Sacramento
CHARLES T. KRESGE, Dow Chemical Company, Midland, MI
SCOTT J. MILLER, Yale University, New Haven, CT
DONALD PROSNITZ, RAND Corporation, Walnut Creek, CA
MARK A. RATNER, Northwestern University, Evanston, IL
ERIK J. SORENSEN, Princeton University, NJ
WILLIAM C. TROGLER, University of California, San Diego, San Diego
THOMAS H. UPTON, ExxonMobil Chemical Company, Baytown, TX

National Research Council Staff

DOROTHY ZOLANDZ, Director
ANDREW C. CROWTHER, Postdoctoral Research Associate
KATHRYN J. HUGHES, Program Officer
TINA M. MASCIANGIOLI, Senior Program Officer
KELA L. MASTERS, Program Associate (through October 2008)
ERICKA M. MCGOWAN, Associate Program Officer
JESSICA L. PULLEN, Administrative Assistant
SHEENA F. SIDDIQUI, Research Assistant
LYNELLE C. VIDALE, Program Assistant

Acknowledgment of Reviewers

This report has been reviewed in draft form by persons chosen for their diverse perspectives and technical expertise in accordance with procedures approved by the National Research Council's Report Review Committee. The purpose of this independent review is to provide candid and critical comments that will assist the institution in making the published report as sound as possible and to ensure that it meets institutional standards of objectivity, evidence, and responsiveness to the study charge. The review comments and draft manuscript remain confidential to protect the integrity of the deliberative process. We wish to thank the following for their review of this report:

Dr. Frances Arnold, California Institute of Technology, Pasadena
Dr. Andreja Bakac, Ames Lab/Iowa State University, Ames
Dr. Avelino Corma, Instituto de Tecnologia Quimica, Valencia, Spain
Dr. Frank DiSalvo, Cornell University, Ithaca, NY
Dr. Cynthia Friend, Harvard University, Cambridge, MA
Dr. Carol Handwerker, Purdue University, West Lafayette, IN
Dr. Scott Miller, Yale University, New Haven, CT
Dr. C. Bradley Moore, University of California, Berkeley
Dr. Hans Niemantsverdriet, Technische Universiteit Eindhoven, Netherlands
Dr. Guido Pez, Air Products, Allentown, PA
Dr. Stuart Soled, ExxonMobil, Annandale, NJ
Dr. Joseph Zoeller, Eastman Chemical Co., Kingsport, TN

Although the reviewers listed above provided many constructive comments and suggestions, they were not asked to endorse the conclusions or recommendations, nor did they see the final draft of the report before its release. The review of this report was overseen by Dr. Marye Anne Fox, University of California, San Diego and Dr. Carl Lineberger, University of Colorado, Boulder. Appointed by the National Research Council, they were responsible for making certain that an independent examination of this report was carried out in accordance with institutional procedures and that all review comments were carefully considered. Responsibility for the final content of this report rests entirely with the authors and the institution.

Contents

Summary		**1**
1.	**Introduction**	**13**
	Catalysis and Chemical Transformations, 13	
	Catalysis and Energy, 15	
	Catalysis and the Department of Energy Mission, 16	
	Catalysis and Basic Research Funding, 17	
	Summary, 19	
2.	**Overview of the Catalysis Science Program**	**21**
	History, 21	
	Budget, 22	
	Current Status, 24	
	Summary, 29	
3.	**Overview of the Catalysis Science Program Portfolio**	**31**
	Research Grants, 32	
	Principal Investigators, 34	
	Conclusions, 41	
4.	**Key Influences on the Development of the Office of Basic Energy Sciences Catalysis Science Program Portfolio**	**43**
	Workshops and Professional Meetings, 43	
	Research Solicitations, 47	
	Contractor Meetings, 48	
	Office of Basic Energy Sciences Advisory Committee, 51	
	Conclusions, 53	
5.	**Analysis of Catalysis Science Program Portfolio**	**55**
	Metrics for Evaluating the Catalysis Science Program Portfolio, 55	
	Heterogeneous Catalysis, 56	
	Homogeneous Catalysis, 74	
	Summary, 88	
6.	**Findings and Recommendations**	**89**
	Findings, 89	
	Recommendations, 93	

Appendixes

A	Statement of Task	95
B	Guest Speaker and Committee Member Biographic Information	97
C	Guest Speakers	109
D	Industry Questionnaire Respondents	111
E	2005 Committee of Visitors Review Excerpt	113
F	Catalysis Science Program Principal Investigators	117

Summary

This report presents an in-depth analysis of the investment in catalysis basic research by the U.S. Department of Energy (DOE) Office of Basic Energy Sciences (BES) Catalysis Science Program.[1] Catalysis is essential to our ability to control chemical reactions, including those involved in energy transformations.[2] Catalysis is therefore integral to current and future energy solutions, such as the environmentally benign use of hydrocarbons and new energy sources (such as biomass and solar energy) and new efficient energy systems (such as fuel cells).[3] On the basis of the information that was evaluated for the preparation of this report, the Committee on the Review of the Basic Energy Sciences Catalysis Science Program (the committee) concludes that BES has done well with its investment in catalysis basic research. Its investment has led to a greater understanding of the fundamental catalytic processes that underlie energy applications, and it has contributed to meeting long-term national energy goals by focusing research on catalytic processes that reduce energy consumption or use alternative energy sources. In some areas the impact of the research has been dramatic, while in others, important advances in catalysis science are yet to be made.

OVERVIEW

Energy (production, storage, and utilization) constitutes one of the most important and challenging issues in the United States. To achieve its mission to advance the national, economic, and energy security of the United States, DOE

---

[1] For the purposes of the report, the DOE's basic research in the science of catalysis is defined by the portfolio of grants that are funded by the BES Catalysis Science Program (formerly the Catalysis and Chemical Transformations Program).

[2] National Research Council. 1992. *Catalysis Looks to the Future.* Washington: National Academy Press.

[3] Basic Research Needs: Catalysis for Energy. U.S. Department of Energy Basic Energy Sciences Workshop. *http://www.sc.doe.gov/bes/reports/files/CAT_rpt.pdf* . Accessed January 30, 2009.

supports basic physical-science research that focuses on energy-related issues.[4] The study of catalysis, the process by which a substance (a catalyst) increases the rate of a chemical reaction, is an important part of the research portfolio. This is because catalysts are essential to energy: they are crucial to the development of new energy technologies and to the processing and manufacturing of fuels for energy storage.[5]

Since 1999, the Catalysis Science Program has sponsored more than 1,000 catalysis basic research grants at universities and national laboratories (Figure S-1). National laboratories have received a smaller number of grants, but the dollar amount of the grants has been split evenly between national laboratories and universities. For fiscal year (FY) 2007, the program was funded at approximately $38 million (3 percent of the BES budget).[6]

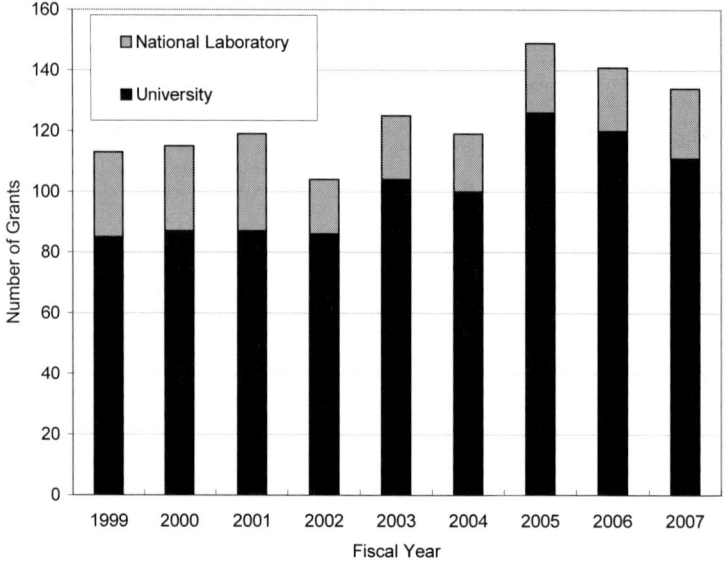

FIGURE S-1 Catalysis basic research grants funded by DOE, FYs 1999–2007. SOURCE: U.S. Department of Energy, Office of Basic Energy Sciences, Catalysis Science Program.

---

[4]About DOE. U.S. Department of Energy. *http://www.doe.gov/about/index.htm*. Accessed May 9, 2008.
[5]Industrial Technologies Program: Chemicals Industry of the Future. U.S. Department of Energy. *http://www1.eere.energy.gov/industry/chemicals/*. Accessed May 9, 2008.
[6]FY 2009 Congressional Budget Request-- Budget Highlights. U.S. Department of Energy *http://www.cfo.doe.gov/budget/09budget/Content/Highlights/Highlight2009.pdf*. Accessed May 9, 2008.

SUMMARY

The grants in the Catalysis Science Program portfolio are distributed among individual researchers and small groups and cover a variety of research areas. Grants for research in heterogeneous catalysis (multiphase reactions catalyzed by solid-state catalysts) include nanoscience, surface science, and theory and have averaged 70 percent of the program's portfolio. Grants for research in homogeneous catalysis (single-phase reactions catalyzed by molecular catalysts) include biocatalysis and have averaged 30 percent of the portfolio (Table S-1).

The program currently funds 170 principal investigators in 78 institutions.[7] According to demographic information collected by the committee, the typical principal investigator being funded during fiscal years 1999 to 2007 was a full professor who had received a Ph.D. approximately 20 years earlier.

TABLE S-1 Funding and Number of Catalysis Science Program Grants for Research in Homogeneous and Heterogeneous Catalysis, FYs 1999–2001, 2002–2004, 2005–2007

		Funding (millions of dollars)	Percentage of Total Funding	No. Grants
Heterogeneous catalysis				
	1999–2001	48	67%	88
	2002–2004	67	74%	102
	2005–2007	79	73%	137
Homogeneous catalysis				
	1999–2001	24	33%	50
	2002–2004	24	26%	39
	2005–2007	30	27%	62

NOTES: Heterogeneous-catalysis grants are focused on multiphase reactions catalyzed by solid-state catalysts and include catalysis science, nanoscience, surface science, theory, and grants under other initiatives. Homogeneous-catalysis grants are focused on single-phase reactions catalyzed by molecular catalysts and include biorelated catalysis.
SOURCE: U.S. Department of Energy, Office of Basic Energy Sciences, Catalysis Science Program.

---

[7] See Appendix F for a list of the names and affiliations of principal investigators who are being funded by the program in FY 2008.

In 2005, Congress passed the Energy Policy Act, which instructed DOE to ask the National Academies to review the Catalysis Science Program (H.R. 6, SEC. 973). BES then called upon the National Academies in 2007 to perform the following tasks:

- Examine the BES research portfolio in catalysis and identify whether and how it has advanced fundamental science.
- Discuss how the BES research portfolio in catalysis contributes and is likely to contribute to meeting immediate and long-term national energy goals, such as reducing the nation's dependence on foreign sources of energy.

A committee of experts in heterogeneous catalysis, homogeneous catalysis, biocatalysis, photocatalysis, surface science, and materials science was convened. The committee, overseen by the Board on Chemical Sciences and Technology, held three meetings during which data from briefings (see Appendix C) and literature reviews were evaluated to determine the committee's findings and recommendations.

Impact of the Catalysis Science Program on Fundamental Science and Future Contributions to National Energy Goals

For the purposes of this study, the committee defined the fundamental science of catalysis as the general understanding of or insight into a catalysis system or a material that is fundamental enough to be applied to more than one specific catalyst. Examples of fundamental science include the development of quantitative models of a class of reactions (such as hydrocarbon oxidation) on a class of catalysts (such as noble metals), the synthesis of a new class of materials (such as zeolites), or the understanding of the reaction or surface mechanisms of a class of catalysts (such as transition metal oxides). Additionally, the committee identified the national energy goals to be the improved production and use of current and future energy sources.

Highlights

Examples of areas where the Catalysis Science Program has had a significant impact on fundamental science or has made contributions to meeting national energy goals are provided below and in Chapter 5.

Modeling Catalytic Structures and Their Reaction Environment

By funding the development of computation methods for the analysis of heterogeneous catalysts for the past 10 years, the Catalysis Science Program has been the main contributor to the growth of theoretical understanding and modeling of surface catalytic structures. As a result, it is now possible to calculate activation energies of elementary surface reactions for various reactions and catalysts and to understand the trends in reactivity from one catalyst to the next. For instance, see the work of the Barteau,[8] Mavrikakis,[9] and Neurock[10].

Nanostructured Metal Oxides

During the past few years, catalysis scientists have dramatically improved their ability to design and synthesize inorganic sites with controlled size, atomic connectivity, and hybridization with either organic or other inorganic superstructures. The resulting materials contain chemical functions and physical properties that can be tuned for energy conversion, petrochemical synthesis, and environmental reactions. Several of the groups that have been funded by the Catalysis Science Program have made contributions to this area of fundamental science, such as Bell,[11] Guliants,[12] Hrbek,[13] Iglesia,[14] Peden,[15] Suib,[16] and Wachs.[17]

---

[8]Linic, S., J. Jankowiak, and M. A. Barteau. 2004. Selectivity driven design of bimetallic ethylene epoxidation catalysts from first principles. J. Catal. 224:489-493.

[9]Zhang, J., M. Vukmirovic, Y. Xu, M. Mavrikakis, R.R. Adzic. 2005. Controlling the catalytic activity of platinum-monolayer electrocatalysts for oxygen reduction with different substrates. Angew. Chem. Int. Ed. 44:2132-2135.

[10]Pallassana, V., and M. Neurock. 2000. Electronic factors governing ethylene hydrogenation and dehydrogenation activity of pseudomorphic PdML/Re(0001), PdML/Ru(0001), Pd(111), and PdML/Au(111) surfaces. J. Catal. 191:301-317.

[11]Rhodes, M.J., and A.T. Bell. 2005. The Effects of Zirconia Morphology on Methanol Synthesis from CO and H2 over Cu/ZrO2 Catalysts: Part II – Transient Response Infrared Studies J. Catal. 233:210-220.

[12]Guliants, V. V., M. A. Carreon, Y. S. Lin. 2004. Ordered mesoporous and macroporous inorganic films and membranes. J. Membr. Sci. 235(1-2):53-72.

[13]Rodriguez, J.A., S. Ma, P. Liu, J. Hrbek, J. Evans, M. Perez. 2007. Activity of CeOx and TiOx nanoparticles grown on Au(111) in the water-gas shift reaction. Science 318(5857):1757-1760.

[14]Liu, H. and E. Iglesia. 2005. Selective oxidation of methanol and ethanol on supported ruthenium oxide clusters at low temperatures. J. Phys. Chem. B 109(6):2155-2163.

[15]Herrera, J.E., J.H. Kwak, J.Z. Hu, Y. Wang, C.H.F. Peden, J. Macht, and E. Iglesia. 2006. Synthesis, characterization, and catalytic function of novel highly dispersed tungsten oxide catalysts on mesoporous silica. J. Catal. 239:200-211.

[16]Yuan, J.K., W.N. Li, S. Gomez, S.L. Suib. 2005. Shape-controlled synthesis of manganese oxide octahedral molecular sieve three-dimensional nanostructures J. Am. Chem. Soc. 127(41):14184-14185.

[17]Wachs, I.E., Y. Chen, J.M. Jehng, L.E. Briand, T. Tanaka. 2003. Molecular structure and reactivity of the Group V metal oxides. Catal. Today 78(1-4):13-24.

Conversion of Biomass for Energy Purposes

The potential of biomass as an alternative source of energy has opened up a field of research that may have a substantial impact on the advancement of science and on progress toward meeting the nation's energy goals. Within the program's grant portfolio, Dumesic and colleagues have made a number of discoveries that were inspired by initial work that deal with the selectivity for cleavage of C-O versus C-C bonds in oxygenated hydrocarbon intermediates on metal surfaces.[18] This work has led to the recent success of a promising biorefinery concept.[19]

Single-Site Polymerization

Catalysis is linked to energy with respect to the large amount of fossil fuel that is consumed by the chemical industry, which DOE estimates to be almost 30 percent of all U.S industrial energy consumption.[20] Fundamental research on ligand design and mechanistic studies by Bercaw and colleagues,[21] and expanded upon by other researchers,[22] has resulted in the development of highly active single-site polymerization catalysts that are now used by U.S. industry to produce over 2 billion pounds of polyolefins a year. The new polymerization processes are more efficient, use less energy, and require less capital than prior technology, which has impacted polymer production around the world.[23]

FINDINGS AND RECOMMENDATIONS

After careful review of the research portfolio (grant titles, abstracts, individual researchers), especially for the fiscal years 1999 to 2007, the committee concludes that BES has done well with its investment in catalysis basic research through the Catalysis Science Program. The program's success can be attributed to key management decisions over the past eight years that have led to a current

---

[18]Cortright, R. D., R. R. Davda, and J. A. Dumesic. 2002. Hydrogen from catalysis reforming of biomass-derived hydrocarbons in liquid water. *Nature* 418:964-967.

[19]Cho, A. 2007. James Dumesic Profile: Catalyzing the emergence of a practical biorefinery. *Science* 315:795.

[20]Industrial Technologies Program: Chemicals Industry of the Future. U.S. Department of Energy. http://www1.eere.energy.gov/industry/chemicals/. Accessed May 9, 2008.

[21]Shapiro, P.J., E. Bunel, W.P. Schaefer, and J. E. Bercaw. 1990. Scandium complex [{(.eta.5-C5Me4)Me2Si(.eta.1-NCMe3)}(PMe3)ScH]2: A unique example of a single-component .alpha.-olefin polymerization catalyst. *Organometallics* 9:867-869.

[22]McKnight, A. L., and R. M. Waymouth. 1998. Group 4 ansa-cyclopentadienyl-amido catalysts for olefin polymerization. *Chem. Rev.* 98:2587-2598.

[23]Chum, P. S., W. J. Kruper, and M. J. Guest. 2000. *Adv. Mater.* 12:1759-1767.

funding distribution that advances catalysis science in general and keeps the development of energy-related technologies as a long-term goal. The program has maintained support for many well-established and world-renowned leaders in catalysis and, at the same time, has brought in many new researchers. The DOE Catalysis Science Initiative (CSI) has been a particularly effective mechanism for bringing to the program new funds, new researchers, and innovative research topics—especially in heterogeneous catalysis. However, there are variations in the quality and relevance of the research in the program's portfolio, as summarized in the committee's main findings and recommendations below.

FINDINGS

The Catalysis Science Program portfolio is distributed between the two main types of catalysis: heterogeneous and homogenous, each of which is assessed separately below. The committee has made this distinction for convenience, based on the traditional division in the science. However, researchers are increasingly crossing the traditional barriers between heterogeneous and heterogeneous catalysis, blurring the distinction between the two (see the discussion on Contractor Meetings in Chapter 4), which the committee views as a definite positive development.

Heterogeneous Catalysis

Research in heterogeneous catalysis accounts for the largest portion of the program's portfolio. For the past eight years (FY 1999–FY 2007), the program has made substantial progress in its support of the experimental and theoretical understanding of multiphase (heterogeneous) catalytic systems, surfaces, and nanoscale structures. Contributions of the portfolio to national energy goals are also discussed where appropriate.

Traditional Heterogeneous Catalysis grants are awarded to individual investigators. These grants have been indispensable in establishing a long-term funding basis for several leading U.S. researchers in the field. The portfolio is highly important to research on the energy efficiency of current chemical transformation processes and on alternative energy solutions. Pioneering work has been conducted in the areas of short-residence-time reactors; basic and acidic properties of catalysts using various probes and spectroscopic techniques; and aqueous-phase reforming of biomass for energy purposes.

Surface Science grants focus on achieving a better understanding of heterogeneous catalytic surfaces. Since its inception, the Catalysis Science Program has supported U.S. leaders in surface science and is now seeing a second generation of principal investigators, many of whom are graduate and postdoctoral students of the science's pioneers. During the past decade, the principal investigators have made numerous contributions to the mechanistic and structural understanding of catalytic reactions, which continue to advance catalysis of energy processes. Examples of this work include hydrogenation and dehydrogenation, reforming, selective oxidation, heteroatom removal, surface photochemistry and catalysis, structure and dynamics of catalyst surfaces, and bimetallic and alloy systems. The work is the foundation of the grand challenge to "Understand the Mechanisms and Dynamics of Catalyzed Transformations," which is articulated in the recent report of the DOE *Basic Research Needs in Catalysis for Energy* workshop.[24]

Research and researchers funded by surface science grants also have contributed substantially to the growth of nanoscience and theory. Historically much of heterogeneous catalysis and the research supporting it have been at the nanoscale. However, the increased and broader focus on nanoscience at the national level has changed the emphasis in surface science. During the most recent three-year time period, approximately one-half of the projects focused primarily on surface reaction mechanisms, and the other half focused more on surface structure.

Nanoscience grants focus on emergent catalytic properties at the nanometer scale. Funding for these grants began in 2001 as a result of the National Nanotechnology Initiative (NNI).[25] Most of the NNI-funded work concentrates on the synthesis of novel single-site heterogeneous catalysts, nanoparticle catalysts, or new materials that might lead to a new family of catalysts. New materials are explored through new synthesis schemes that are used to make catalytic porous solids or by incorporating catalytic species into solid supports. Ten awards were originally funded under the NNI, and seven of them were still being funded in 2007. Overall, the recent influx of funding for the Catalysis Science Program under the NNI has led to funding of several new investigators.

Catalysis Science Initiative (CSI) grants were first awarded in 2003 and were given to multi-investigator, multidisciplinary teams mainly involved in heterogeneous catalysis research. Few grants have been awarded for research in homogeneous catalysis or biocatalysis, despite the initiative's broader goal to develop "combined experimental and theoretical approaches to enable molecular-level understanding of catalytic reaction mechanisms." Although the 11 programs currently funded by the CSI are less than six years old, they already represent approximately 20 percent of the heterogeneous catalysis portfolio and

---

[24]Basic Research Needs: Catalysis for Energy. U.S. Department of Energy Basic Energy Sciences Workshop. *http://www.sc.doe.gov/bes/reports/files/CAT_rpt.pdf*. Accessed January 30, 2009.

[25] The NNI is an interagency program to coordinate federal nanotechnology research and development. Investments in NNI are overseen by the President's Council of Advisors on Science and Technology and the program is reviewed every three years by the National Research Council.

have been successful in attracting and supporting investigators new to the field. This record suggests that the CSI has added value to the Catalysis Science Program and has advanced the field of catalysis.

Theory grants focus on theory, modeling, and simulation. Grants in other categories include theory but not as a main focus. Because the field is new, several grants have been used to build programs. The catalysis theory portfolio is considered to be of a high international standard. The list of grantees includes most of the leading U.S. researchers in the field. However, the current portfolio is somewhat lacking in the development of theoretical and computational methods, as well as in work focused on homogeneous catalysis and biocatalysis.

Hydrogen Fuel Initiative (HFI) grants focus on hydrogen production, storage, and use and mainly involve electrocatalysis. Many of the HFI-funded projects study the fundamental aspects of catalysis related to specific applications, such as catalysis for fuel cells or for reforming. Because funding began in FY 2005 for most electrocatalysis projects and in FY 2007 for other projects, it is difficult to assess the impact of this body of work. However, the collection of electrocatalysis and catalysis research in the portfolio is appropriate. The research mostly reflects the technical challenges that arise when fuel hydrogen is produced from hydrocarbon resources (for example, carbon monoxide poisoning on platinum electrodes and the use of catalysts for reforming methane) rather than from electrolysis of water by solar or nuclear means. In addition, and similar to the CSI, these new HFI-funded projects have attracted new researchers to the Catalysis Science Program.

Homogeneous Catalysis

Grants for research in homogeneous catalysis constitute a smaller portion of the current portfolio but have had an important impact on the Catalysis Science Program. For FY 2007, the grants were divided into two main research topics: approximately one-half involved C-H activation, and the other half involved inorganic synthesis (including inorganic single sites and polymerization). The committee also assessed the research topics of homogeneous catalysis in organic synthesis and in biorelated projects.

Single-Site Polymerization grants have made significant contributions to the understanding of fundamental catalysis. Single-site polymerization is one of the important advances in catalysis of the past 25 years. The Catalysis Science Program has strongly supported single-site polymerization research from the inception of the field and must be credited with having a great impact on its development. This is an excellent example of the value of basic research and of how funding productive, well-qualified individual principal investigators can lead to a successful commercial result of huge importance to chemical production and energy utilization.

C-H Activation and Functionalization grants have been a part of the Catalysis Science Program for a long time. The program has made major contributions to successes in fundamental research in this area. The ultimate goal of research in C-H activation catalysis is to find catalysts that will incorporate C-H activation into hydrocarbon-conversion technology, which will lead to functionalized compounds needed for feedstocks in the chemical industry or to the conversion of methane into useful liquid transportation fuels. However, the program has limited its impact by focusing its support on studies of C-H activation. Simple functionalization of hydrocarbons after C-H activation has not been realized, and new ideas are needed. Designs based on alkyl group transfer to a second metal or on bifunctional ligands are possibilities. The study of C-H functionalization in biological processes also could help to inform research in this area.

Homogeneous Catalysis in Organic Synthesis grants are a very small but still important part of the Catalysis Science Program portfolio. For example, the high inherent selectivity of homogeneous catalysts allows the production of molecules of desired handedness or enantioselectivity (asymmetric catalysis), which is critical for the synthesis of fine chemicals, pharmaceuticals, agricultural chemicals, and electronic material. The selectivity of these catalysts presents the potential to conserve resources, increase energy efficiency, and reduce waste.

Biorelated grants are another small but important part of the Catalysis Science Program portfolio. Biological processes provide understanding of important catalytic reactions such as C-H functionalization. Many projects in the homogeneous catalysis portfolio are described as bioinspired, but there are only a few examples of research that carefully analyzes the mechanistic implications of enzyme active sites and the requirements met by the surrounding protein matrix. Several of the program's principal investigators are active in bioinorganic chemistry but receive support for that work from government agencies other than DOE.

RECOMMENDATIONS

The Catalysis Science Program should continue its current approach to funding decisions. Multi-investigator and interdisciplinary programs such as the Catalysis Science Initiative should remain a part of the portfolio, but future teams might benefit from the inclusion of more homogeneous catalysis and biocatalysis researchers that are interested in energy solutions. The program should utilize future funding initiatives as a mechanism to maintain the balance of experienced and new researchers in the program and to explore new approaches to carrying out research.

Influences on the Portfolio

The Catalysis Science Program should continue to broaden participation in its contractor meetings and other activities. Non-DOE sponsored workshop organizers and grantees funded by other Office of Basic Energy Sciences programs should be invited to attend the Catalysis Science Program's activities to provide a more diverse influence on the portfolio. This is particularly important in the development of research directions that will have a long-term impact on the program.

Principal Investigators

The Catalysis Science Program should continue on its current path of maintaining support for productive, long-term researchers and of recruiting new researchers. The program also must ensure that the best researchers are identified and supported—this is especially important in heterogeneous catalysis, because program funding is essential to the success of a heterogeneous catalysis researcher (see Chapter 3). The balance of funding for individual investigators and small groups should also be maintained.

Heterogeneous Catalysis

The distribution of grants in the heterogeneous catalysis portfolio should be changed slightly. Studies on high surface area catalysts, surface science, nanoscience, and electrocatalysis should be maintained, but there should be a stronger emphasis on studies that explore catalyst design and new synthesis methods, unique reactor systems, unique characterization techniques, and completely new chemical reactions. Support for the development of theoretical methods also should feature more prominently in the portfolio.

Homogeneous Catalysis

A balanced homogeneous catalysis portfolio should extend beyond individual mechanistic steps to include greater development of new catalytic systems and reactions. The portfolio can be improved by pursuing opportunities in C-H bond functionalization, new approaches to transition-metal catalysis, and electrochemical catalysis (small molecule homogeneous catalysts supported on electrodes). In addition, there should be a greater emphasis on reducing highly oxidized compounds such as bioderived materials into fuels and feedstocks, and on bioinspired catalytic processes.

CONCLUSION

The Catalysis Science Program is the primary funder of catalysis basic research in the United States, especially in the area of heterogeneous catalysis. The program has supported many well-established researchers who are world leaders in catalysis science. It has also supported many new researchers, who have largely entered the program through special initiatives, such as the Catalysis Science Initiative and the Hydrogen Fuel Initiative. The program has and should continue to play a key role in meeting national energy needs.

1

Introduction

This report presents an in-depth analysis of the investment in catalysis basic research by the U.S. Department of Energy (DOE) Office of Basic Energy Sciences (BES) Catalysis Science Program during the fiscal years 1999 to 2007. The review examines the BES research portfolio in catalysis and identifies whether and how it has advanced fundamental science and discusses how it contributes and is likely to contribute to immediate and long-term national energy goals, such as reducing the nation's dependence on foreign sources of energy. First, however, it is important to understand what catalysis is and why it is important to chemical transformations, global energy issues, and DOE.

CATALYSIS AND CHEMICAL TRANSFORMATIONS

Catalysis is a process by which a substance (a catalyst) increases the rate of a chemical reaction. Unlike the reactants, the catalyst remains essentially unchanged, relative to its initial state, at the end of the chemical reaction that it facilitates. Catalytic processes are typically categorized as heterogeneous or homogeneous:

- In **heterogeneous catalysis,** the catalyst (typically a solid) is in a different phase from the reactants.
- In **homogeneous catalysis,** the catalyst is in the same phase (typically liquid) as the reactants.

Catalysts vary in composition from solid metal surfaces to enzymes in solution, and they are involved in chemical transformations as different as the refining of petroleum and the synthesis of pharmaceuticals. Catalysis affects almost every aspect of our economy, health, and way of life as documented etensively in the 1992 National Research Council report, *Catalysis Looks to the Future.*[1]

Catalysis is especially critical in the chemical and petroleum-processing industries, which are the two largest industrial energy users in the United States.[2] The U.S. chemical industry alone is estimated to account for approximately a quarter of global chemical production ($450 billion per year),[3] thus the impact of catalysis on the U.S. economy is substantial. Furthermore, success in catalysis research has contributed to the strong position of the U.S. chemical industry. Examples of industrially important catalysts include

Single-site polymerization catalysts, organometallic-based catalysts used in U.S. industry to produce over 2 billion pounds of polyolefins every year. Some of the polyolefins include long-chain branched copolymers of ethylene with α-olefins, new elastomers, and ones produced as a result of a new process for ethylene propylene diene monomer (EPDM) rubber. The new EPDM polymerization processes are more efficient, use less energy, and use less capital than prior technology. As a result, most of the world production of EPDM materials now uses the single-site polymerization catalysts.

Platinum-group metal catalysts, which have been used in catalytic converters to reduce automobile tailpipe emissions. Catalytic converters have been used on all new cars since the mid 1970s. On the basis of measurements by the Environmental Protection Agency at over 250 sites, the average carbon monoxide (CO) concentration has dropped by 60 percent from 1990 to 2005—largely because of the use of catalytic converters. Most cars today are equipped with three-way catalytic converters, which use newer catalysts that reduce emissions of CO, hydrocarbons or volatile organic compounds, and nitrogen oxides. Unfortunately, the optimal fuel mix for effective catalytic converter operation is often not the same as the optimal fuel efficiency mix, increasing carbon dioxide production.

Zeolite catalysts, crystalline microporous materials that are used in a wide variety of industries, from oil refining to production of fine chemicals.[4] Zeolites are key catalysts in the petroleum refinery units known as fluid catalytic crackers, which are at the heart of gasoline and diesel production. Zeolites have enabled

---

[1]National Research Council. 1992. *Catalysis Looks to the Future.* Washington, DC: National Academy Press.
[2]2002 Manufacturing Energy Data Tables. Energy Information Administration. *http://www.eia.doe.gov/emeu/mecs/mecs2002/data02/shelltables.html*. Accessed February 2, 2009.
[3]Industrial Technologies Program: Chemicals Industry of the Future. U.S. Department of Energy. *http://www1.eere.energy.gov/industry/chemicals/.* Accessed May 9, 2008.
[4]Davis, M.E. 2003. Materials Science: Distinguishing the (Almost) Indistinguishable. *Science* 300(5618):438-439.

the petroleum industry to increase the gasoline obtained from a barrel of oil (1 barrel = 42 gal) from 14 gal in the 1960s to 20 gal in the late 1980s.[5]

CATALYSIS AND ENERGY

For fossil fuels (petroleum, natural gas, and coal) to be used with energy technologies (such as internal-combustion engines in automobiles), raw materials must be chemically modified. In the case of petroleum, for example, the petroleum-refining industry (much like the chemical industry) relies heavily on catalysis to achieve the desired chemical form of the desired final product. Beyond the manufacture of the fuel, catalysis plays a critical role in mitigating the impact of the use of fossil fuel, as illustrated by the dramatic improvement in urban air quality as a result of using platinum-group catalysts in catalytic converters. There is no doubt that catalysis plays a critical role in today's energy technologies.

Catalysis is also linked to energy with respect to the large amount of fossil fuel consumed by the chemical industry, estimated by DOE to be almost 30 percent of all U.S. industrial energy consumption.[6] Over half of that "energy" is fuel, such as natural gas, used as chemical feedstocks for the manufacture of more valuable products, and the rest is consumed primarily to generate electricity and heat for manufacturing processes. More efficient catalytic processes have the potential to decrease fossil-fuel consumption in both feedstocks and process heat. That is, catalysis that produces the desired end products at higher yields reduces consumption of the raw materials, and catalysis that increases reaction rates reduces the heat required to drive a process.

Catalysis will be critical for converting other resources such as biomass and sunlight to usable sources of energy, and for developing new efficient processes for using them in fuel cells and other new technologies. Catalysis is thus fundamentally linked to both energy delivery and energy use. The importance of catalysis related to energy is extensively covered in the 2008 DOE report, *Basic Research Needs: Catalysis for Energy*,[7] and underlies DOE's financial support for research in catalysis, which will be discussed in more detail below.

---

[5]Katz, R. N. 2001. Advanced ceramics: Zeolites = More miles per barrel. *Ceramic Industry Magazine*. http://www.ceramicindustry.com/. Accessed January 31, 2009.
[6]Industrial Technologies Program: Chemicals Industry of the Future. U.S. Department of Energy. http://www1.eere.energy.gov/industry/chemicals/. Accessed May 9, 2008.
[7]Basic Research Needs: Catalysis for Energy. U.S. Department of Energy Basic Energy Sciences Workshop. http://www.sc.doe.gov/bes/reports/files/CAT_rpt.pdf. Accessed January 31, 2009.

CATALYSIS AND THE DEPARTMENT OF ENERGY MISSION

DOE's overarching mission is to advance the national, economic, and energy security of the United States. The agency also seeks to promote scientific and technologic innovation in support of that mission and to ensure the environmental cleanup of the national nuclear-weapons complex.[8] Catalysis science is relevant to the mission of DOE and the welfare of the nation in that, as stated earlier, catalysts are needed for the processes that convert crude oil, natural gas, coal, and biomass into clean-burning fuels; catalysts are crucial for energy conservation in creating new, less energy-demanding routes for the production of basic chemical feedstocks and value-added chemicals; and catalysis science has affected the technology that is used to clean up environmental pollutants, such as unwanted emissions from chemical production or combustion, and has provided a means of replacing undesirable chemicals with more benign ones, for example, the displacement of chlorofluorocarbons with more environmentally acceptable refrigerants.[9]

The origin of DOE traces back to the 1940s and the development of nuclear weapons during World War II. However, it was not until 1977, soon after the 1973 oil embargo crisis, that DOE was created "to provide a framework for a comprehensive and balanced national energy plan by coordinating and administering the energy functions of the federal government. The DOE undertook responsibility for long-term, high-risk research and development of energy technology, federal power marketing, energy conservation, the nuclear weapons program, energy regulatory programs, and a central energy data collection and analysis program."[10]

Support for catalysis research at DOE also began at that time as part of the Chemical Energy Program. The Chemical Energy Program encompassed "organic, inorganic, physical and electrochemistry; thermochemistry and reaction mechanisms and dynamics; coal and hydrocarbon fuel chemistry, heterogeneous and homogeneous catalysis, chemistry of hydrogen production and storage, biomass conversions."[11] Most of the support for catalysis basic research at DOE resides in the Catalysis Science Program, but catalysis-related research is also carried out in other programs in BES. Solar photochemistry, energy biosciences, chemical physics, and materials chemistry programs also fund catalysis-related projects, although catalysis does not have the highest priority in these programs.

---

[8]About DOE. U.S. Department of Energy. http://www.doe.gov/about/index.htm. Accessed January 31, 2009.

[9]Based on *CRA-Catalysis Science-2008*, program description, February 2008.

[10]Origins and Evolution of the Department of Energy. U.S. Department of Energy. http://www.doe.gov/about/origins.htm. Accessed July 15, 2008.

[11]U.S. Department of Energy. 1979. DOE/ER-0024, Summaries of FY 1978 Research in the Chemical Sciences, April 1979, National Technical Information Service.

CATALYSIS AND BASIC RESEARCH FUNDING

DOE funding for catalysis basic research is more than that of any other federal agency on the basis of the estimated number of grants that were funded in FY2005 (Table 1-1). In FY2005, DOE funded approximately $35 million in grants, or 54 percent of the total. The National Science Foundation (NSF) and the National Institutes of Health (NIH) also made sizable contributions to catalysis basic research. The three combined provided approximately $65 million. However, the focus of NIH projects is generally much different, that is, targeting synthesis of pharmaceuticals rather than commodity chemicals or energy.

Figure 1-1 shows the decadal trend in the distribution of all catalysis-related basic-research grants by the three main funding agencies (based on information from DOE, the NSF Web site, and the NIH CRISP database). It can be seen that the number of grants for catalysis research has been roughly constant over the past decade.

TABLE 1-1 Estimated U.S. Federal Government Funding for Catalysis Basic Research, FY2005

	No. Grants	% of Grants	Estimated Average Award	Award Duration	Total Funding	% of Total Funding
DOE	151	53%	$235,000	3 years	$35,485,000	54%
NSF	76	27%	$126,000	2–3 years	$9,576,000	15%
NIH	58	20%	$360,000	3–4 years	$20,880,000	32%
Total	285				$65,941,000	

SOURCE: Original analysis based on data from funding agencies. NSF data from Division of Chemical, Bioengineering, Environmental, and Transport Systems and Division of Chemistry collected by searching Awards Database (http://www.nsf.gov/awardsearch/index.jsp); DOE data provided by Office of Basic Energy Sciences; NIH data from CRISP database *(http://crisp.cit.nih.gov)*.

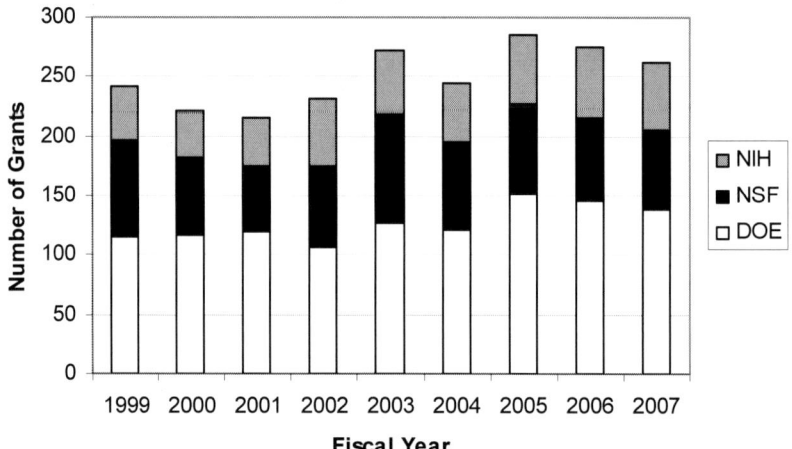

FIGURE 1-1 Distribution of catalysis-related grants by three main government funding agencies.
SOURCE: Original analysis based on data from agencies. NSF data from Division of Chemical, Bioengineering, Environmental, and Transport Systems and Division of Chemistry (http://dellweb.bfa.nsf.gov and http://www.nsf.gov/awardsearch/index.jsp);
DOE data from Office of Basic Energy Sciences; NIH data from CRISP database (http://crisp.cit.nih.gov) and NIH Extramural Data Book, March 2008.

The U.S. federal government investment in catalysis basic research ($65 million, of which $45 million is for nonpharmaceutical research) is similar to the levels of support for individual research institutes in other countries. For example, heterogeneous catalysis research in Europe, Japan, and other Asian countries (on the basis of visits to those countries) was recently summarized as follows:[12]

- Instituto de Tecnologia Quimica, Spain: $6 million budget, 25 research personnel (six principal investigators).
- Fritz-Haber Institute, Germany: $35–40 million budget, 250–300 research personnel.
- Dalian Institute of Chemical Physics, China: $15 million budget, 500 research personnel.

---

[12]Davis, R. 2008. An International Assessment of Research in Catalysis by Nanostructured Materials. Presentation to the Committee on the Review of Basic Energy Sciences Catalysis Program, March 17, 2008.

The funding mechanisms in those countries are quite different from those in the United States. For example, the Fritz-Haber Institute receives 60 percent of its funds from the Max Planck Society, and the Instituto de Tecnologia Quimica receives approximately 80 percent of its budget from sources other than the federal government, including industrial contracts, intellectual property licensing fees, and European projects. In the United States, the main support for catalysis basic research conducted at universities and national laboratories comes from the federal government.

As discussed in the benchmarking analyses conducted by the National Research Council, the maintenance of U.S. leadership and competitiveness in catalysis research may be threatened because of current levels of funding in the United States compared with those in other countries.[13]

SUMMARY

Catalysis—the process by which a substance increases the rate of a chemical reaction—is involved in many chemical transformations of importance to the U.S. economy. The chemical transformations are in turn inextricably linked to production and use of energy. Thus, catalysis plays an essential role in the mission of DOE with respect to both current and long-term national energy goals.

Support for catalysis basic research underlies the fundamental understanding of chemical transformations involved in energy, human health, environmental, and other applications that address societal needs. DOE is the key provider of funding for catalysis basic research with relevance to energy in the United States. The primary mechanism by which DOE supports catalysis basic research is the Catalysis Science Program, which is the main subject of this report.

An overview of the Catalysis Science Program will be provided in Chapter 2, followed by an overview of the research portfolio in Chapter 3, and a discussion of the key influences on the development of the portfolio in Chapter 4. The statement of task will be addressed largely in Chapters 5 and 6 with an in-depth analysis of the portfolio, its impact on fundamental science, and its contributions to reaching national energy goals.

---

[13]National Research Council. 2007. The Future of U.S. Chemistry Research: Benchmarks and Challenges. Washington, DC: National Academies Press; National Research Council. 2007. International Benchmarking of U.S. Chemical Engineering Research Competitiveness. Washington, DC: National Academies Press.

2

Overview of the Catalysis Science Program

This chapter presents an overview of the history, budget, and current status of the Catalysis Science Program, through which the U.S. Department of Energy (DOE) provides support for catalysis basic research.

HISTORY

The Office of Science, one of DOE's eight program offices, is responsible for supporting basic research in the physical sciences (Figure 2-1). Fundamental research is managed by the Office of Science through six interdisciplinary program offices: Advanced Scientific Computing Research, Basic Energy Sciences (BES), Biological and Environmental Research, Fusion Energy Sciences, High Energy Physics, and Nuclear Physics.[1] Of those, BES is responsible for fundamental research in the natural sciences that are applicable to improving energy-related technologies, understanding and mitigating environmental impacts of energy use, and developing the knowledge and tools needed to strengthen national security.[2] BES supports research through its divisions of Materials Sciences and Engineering; Chemical Sciences, Geosciences, and Biosciences (CSGB); and Scientific User Facilities. The Catalysis Science Program is in CSGB, along with five other programs: Atomic, Molecular, and Optical Sciences; Chemical Physics Research; Heavy Element Chemistry; Solar Photochemistry; and Separations and Analysis.

---

[1] About the DOE Office of Science. U.S. Department of Energy. http://www.science.doe.gov/about/. Accessed February 2, 2009.
[2] Office of Basic Energy Sciences. U.S. Department of Energy. http://www.sc.doe.gov/bes/bes.html. Accessed February 2, 2009.

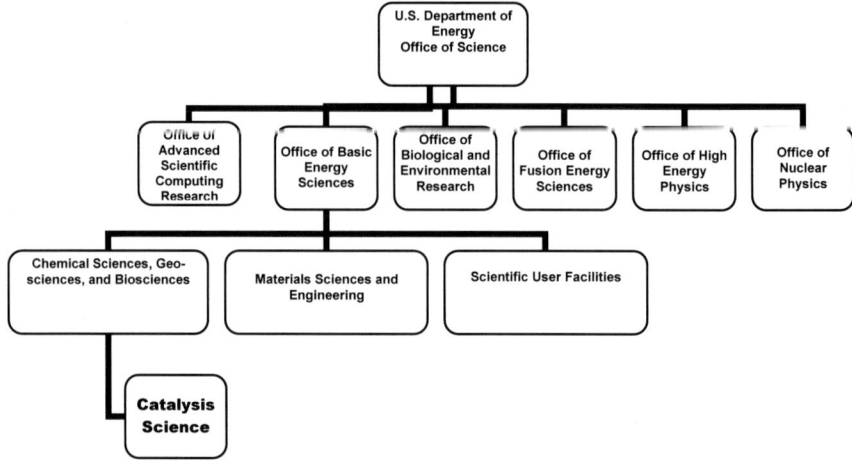

FIGURE 2-1 Organizational structure of U.S. Department of Energy Office of Science.
SOURCE: Adapted from "NAS Review of the BES Catalysis Science Program," presentation by Eric Rohlfing, Office of Basic Energy Sciences (see Appendix C).

BUDGET

As shown above, funding for the Catalysis Science Program originates in congressional appropriations for the Office of Science.[3] In 2007, the Office of Science was allocated $3.8 billion (16 percent of DOE's total budget). Almost one-third of the Office of Science budget is allocated to BES. In fiscal year (FY) 2007, approximately $38 million, or 3 percent, of the BES budget was allocated to the Catalysis Science Program. The budget for catalysis science has increased since 2001 largely because of the Catalysis Science Initiative and new funding for the Hydrogen Fuel Initiative[4] (Figure 2-2).

---

[3]Office of the Chief Financial Officer. U.S. Department of Energy. http://www.cfo.doe.gov/crorg/cf30.htm. Accessed January 9, 2009.
[4]Funds for the Catalysis Science Initiative were made available through restructuring of existing BES budgets and funds. Funds for the Hydrogen Fuel Initiative were newly appropriated and provided in addition to the existing BES budget.

OVERVIEW OF THE CATALYSIS SCIENCE PROGRAM 23

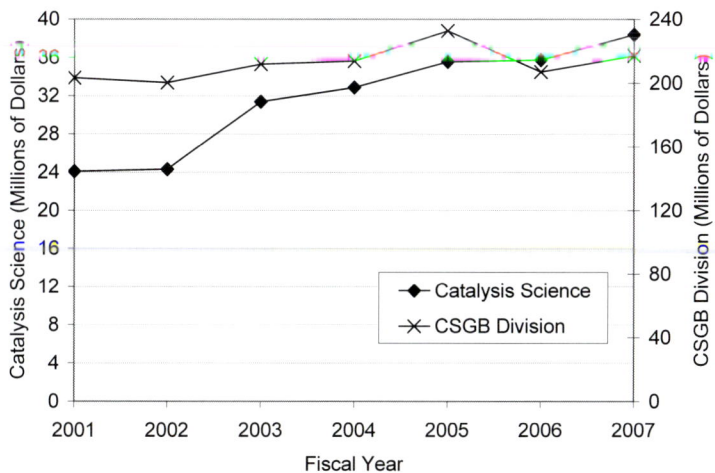

FIGURE 2-2 Comparison of Catalysis Science Program and CSGB funding, FYs 2001–2007.
SOURCE: "NAS Review of the BES Catalysis Science Program," presentation by Eric Rohlfing, Office of Basic Energy Sciences (see Appendix C).

Over the past eight years, three funding initiatives—in nanoscience, hydrogen fuel, and catalysis science—have played an important role in shaping the Catalysis Science Program portfolio. The initiatives are briefly described below and will be discussed in more detail in Chapters 3 and 4.

Nanoscience Initiative

During the Clinton administration (1993–2001), the National Nanotechnology Initiative was created to organize research funding for nanoscale science, engineering, and technology among government agencies.[5] In DOE, research focusing on emergent properties at the nanoscale is spread across BES programs, including the Catalysis Science Program. The total funding for research identified as related to nanoscience in the Catalysis Science Program for FY 1999 to FY 2007 was approximately $37.7 million, which included new money for the program, not just for the reorganization of grants.

---

[5]National Nanotechnology Initiative. *www.nano.gov*. Accessed February 2, 2009.

Hydrogen Fuel Initiative

President George W. Bush, in his 2003 State of the Union address, announced a plan to decrease the nation's dependence on foreign oil by developing hydrogen fuel-cell technology through the Hydrogen Fuel Initiative (HFI). In response to the announcement, DOE sponsored new research and the DOE Hydrogen Program. The HFI was appropriated $1.2 billion by Congress for research related to hydrogen fuel-cell technology. The HFI increased the funds available for research projects throughout DOE (Table 2-1). Since 2004, BES has received more than $130 million for the HFI. The funds support research that focuses on novel materials for hydrogen storage, functional membranes, and nanoscale catalysts. A one-time funding increment of $50 million was allotted to CSGB, including approximately $12 million for the Catalysis Science Program.

Catalysis Science Initiative

The Catalysis Science Initiative was created as a mechanism to encourage "high-risk, long-term, multi-investigator, multidisciplinary research on the science of catalysis."[6] In 2003, 13 new groups were funded with budgets of up to approximately $900,000 per year. Multi-investigator teams were sought, but industry and single investigators were not prevented from competing for Catalysis Science Initiative awards.

CURRENT STATUS

As the largest federal supporter of fundamental research in heterogeneous and homogeneous catalysis in the United States, DOE's Catalysis Science Program funded more than 1,000 research grants in catalysis from FY 1999 to FY 2007 (Figure 2-3). Those grants were provided to individual researchers and small groups in academe and at national laboratories. The program encourages multidisciplinary collaboration between its researchers and supports research that encompasses different types of catalysts, catalytic processes, and techniques, which will be described in more detail below.

In addition to research grants, the Catalysis Science Program supports research centers and workshops as other mechanisms for sharing knowledge and fostering collaboration.

---

[6]U.S. Department of Energy. 2003. Office of Sciences Notice 03-16, Catalysis Science. *http://www.science.doe.gov/grants/Fr03-16.html*. Accessed February 2, 2009.

TABLE 2-1 Hydrogen Fuel Initiative Budget (millions of dollars)

Department/Office	2001[a]	FY 2004[b]	FY 2005[b]	FY 2006[b]	FY 2007[b]	FY 2008	FY 2009 Request
Energy/Energy Efficiency and Renewable Energy	73.0	144.9[c]	166.8[c]	153.5[c]	189.6	211.1	177.7
Energy/Fossil Energy (Coal)	0.0	4.9	16.5[d]	21.0	23.0	24.7	11.4
Energy/Nuclear Energy	0.0	6.2[e]	8.7[e]	24.1[e]	18.8	9.9	16.6
Energy/Basic Energy Sciences	0.0	0.0[f]	29.2	32.5	36.4	36.4	60.4
Transportation	0.0	0.6	0.5	1.4	1.4	1.4	1.4
TOTAL	73.0	156.6	221.7	232.5	269.2	283.5	267.5

[a]Shown for comparison; 2004 was first year for the HFI. Reflects funding for baseline that the HFI augments or redirects.
[b]Reflects rescissions, general reductions, and other adjustments included in relevant appropriations.
[c]Includes $42.0 million in FY 2004, $40.2 million in FY 2005, and $42.5 million in FY 2006 of congressionally directed spending.
[d]Includes $3.0 million in FY 2005 of congressionally directed spending.
[e]Includes $2.0 million in FY 2004, $4.0 million in FY 2005, and $5.0 million in FY 2006 of congressionally directed spending.
[f]Base funding for hydrogen-related activities in BES was roughly $8.0 million in 2004; these activities have been reoriented and expanded to support the goals of the President's HFI in 2005.
SOURCE: DOE Hydrogen Program: Budget. U.S. Department of Energy. http://www.hydrogen.energy.gov/budget.html. Accessed February 5, 2009.

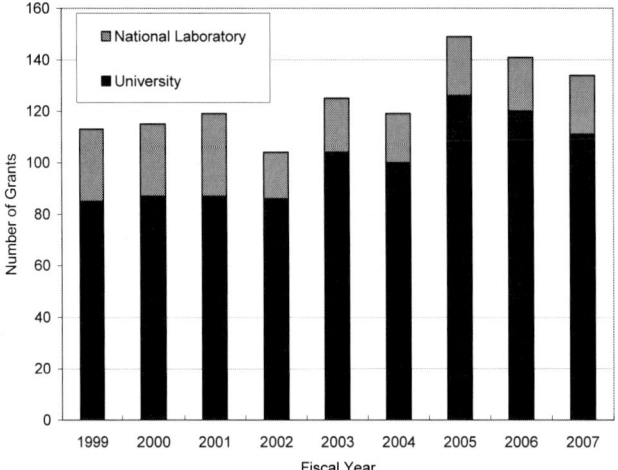

FIGURE 2-3 Catalysis basic research grants funded by DOE, FYs 1999–2007. SOURCE: U.S. Department of Energy, Office of Basic Energy Sciences, Catalysis Science Program.

Programmatic Activities

BES states that it strives to understand how electronic, molecular, and material structures determine reaction mechanisms and kinetics and to control mechanisms and kinetics by means of catalytic structures designed a priori.[7] Support of research grants to university and national laboratories is the primary mechanism for achieving these goals. The program also funds facilities and possibly will fund centers in 2009. Program progress is monitored through contractor meetings and evaluation by the Basic Energy Sciences Advisory Committee (BESAC). These activities are discussed in more detail in Chapter 5.

Research Grants

As mentioned earlier, the Catalysis Science Program has sponsored more than 1,000 research grants at universities and national laboratories since FY 1999 (Figure 2-3). Universities have received a larger number of grants than national laboratories, but the overall dollar amount for grants has been split almost equally between universities and national laboratories. For example, the average grant size in FY 2005 was approximately $235,000 per year for three years (see Table 1-1). However, the average grant size for universities was approximately $140,000 per year, and the average for national laboratories was approximately $700,000 per year.

---

[7] "Catalysis Science Program: Chemical Transformations Team," presentation to the Committee on the Review of the Basic Energy Sciences Catalysis Science Program, January 10, 2008 by Raul Miranda, Office of Basic Energy Sciences (see Appendix C).

Projects funded by the Catalysis Science Program are distributed among the following areas:

- Homogeneous catalysis
- Heterogeneous catalysis
- Surface science
- Nanoscience
- Catalysis Science Initiative
- Biorelated catalysis
- Theory
- Hydrogen Fuel Initiative

These areas overlap in that more than one aspect of catalysis may be needed to address a single research question. DOE recognizes the importance of multidisciplinary collaboration to address modern catalysis-related issues, but grants typically are provided to individual investigators. The Catalysis Science Program staff says that it does not have a fixed target allocation of single-investigator versus multi-investigator projects; however, multi-investigator and multidisciplinary collaboration is encouraged and is often specified in requests for applications.

Facilities

In fulfillment of its mission, BES plans, constructs, and operates user facilities that are available for academic, national laboratory, and industrial scientists.[8] The facilities provide specialized instrumentation and expertise that are not available in the researchers' own laboratories. The Synchrotron Catalysis Consortium is a group of academic, national laboratory, and industrial institutions specifically funded by the Catalysis Science Program. It leverages resources at the National Synchrotron Light Source, specifically two beamlines, and promotes "the utilization of synchrotron techniques to perform cutting-edge catalysis nano-science research under in-situ conditions." The consortium consists of principal investigators and others from academic, national, and industrial laboratories who have extensive catalysis expertise. It provides expert staff, training courses, and facilities in an effort to assist in the development of science and techniques in the catalysis community.[9]

Research Centers

In an effort to guide research, BES sponsored 12 workshops that attracted participants from academe, industry, and national laboratories. In re-

---

[8]BES Scientific User Facilities. Office of Basic Energy Sciences. *http://www.er.doe.gov/bes/BESfacilities.htm*. Accessed February 2, 2009.

[9]Synchroton Catalysis Consortium. Yeshiva University. *http://www.yu.edu/scc/*. Accessed February 2, 2009.

sponse to recommendations resulting from the workshops, BES proposed a two-pronged approach to fostering multidisciplinary collaboration to address critical scientific challenges related to energy: multi-investigator research in Energy Frontier Research Centers (EFRCs)[10] and enhancement of Single-Investigator and Small-Group (SISGR)[11] projects that now form the bulk of the BES core research portfolio. DOE posted a funding opportunity announcement for the first EFRCs in early spring 2008 and expects to begin distributing EFRC awards of $2–5 million per year for five years beginning in FY 2009 ($100 million in new funds anticipated in the FY 2009 budget). Approximately $60 million also will be available for SISGR awards; the initial award period is expected to be three years. Single-investigator awards are expected to be for approximately $150,000–300,000 per year, and small-group awards for $0.5–1.5 million per year.

In addition, the Catalysis Science Program is using the Nanoscale Science Research Centers (NSRCs), which are national user facilities created by BES to support science in nanomaterials and nanosystems, instrumentation, and theory. Two of the NSRCs—at Oak Ridge National Laboratory and at Brookhaven National Laboratory—focus on catalysis. The Argonne National Laboratory NRSC also has substantial core strength in catalysis.

Contractor Meetings

Since 1999, the Catalysis Science Program has conducted annual contractor meetings as a means of sharing information and encouraging contact among and within disciplines. Contractor meetings differ from regular professional conferences. These meetings are designed to provide a sense of partnership among principal investigators by having them identify future directions for their core funding program (such as the Catalysis Science Program). Therefore, the contractor meetings are limited to current principal investigators and selected outside speakers. The meetings are intended to foster in-depth discussion of recent research results and of needs and opportunities for the program. Contractor meetings are discussed in more detail in Chapter 4.

Basic Energy Sciences Advisory Committee

The BES Advisory Committee (BESAC) is charged with providing independent advice to DOE on the complex scientific and technical issues that arise in the planning, management, and implementation of the BES program.[12] BESAC periodically reviews BES program elements and provides guidance on program directions, priorities, and funding. BESAC was involved in spawning

---

[10]Energy Frontier Research Centers. U.S. Department of Energy Basic Energy Sciences. http://www.science.doe.gov/bes/EFRC.html. Accessed December 19, 2008.

[11]Single-Investigator and Small- Group Research. U.S. Department of Energy. http://www.er.doe.gov/bes/SISGR.html. Accessed December 19, 2008.

[12]Basic Energy Sciences Advisory Committee (BESAC). Department of Energy. http://www.sc.doe.gov/bes/BESAC/BESAC.htm. Accessed February 2, 2009.

the 12 BES workshops that led to the development of EFRCs. It consists of scientists in the academic, industrial, and national laboratory communities.

SUMMARY

Catalysis research at DOE has its roots in the Chemical Energy Program, which existed in CSGB from 1977 to 1999. Today, the Catalysis Science Program essentially maintains the mission of the Chemical Energy Program but has been refined to focus on basic research to understand the chemical aspects of catalysis, both heterogeneous and homogeneous; the chemistry of fossil resources; and the chemistry of the molecules used to create advanced materials.[13] In 2007, the Catalysis Science Program budget was approximately $38 million, up from $24 million in 2001 (Figure 2-4). The increase was due to funding from Catalysis Science Initiative in 2003 and the HFI in 2005 and 2007.[14]

The Catalysis Science Program is the largest federal supporter of fundamental research in heterogeneous and homogenous catalysis in the United States and has sponsored research activities through more than 1,000 research grants in universities and national laboratories since FY 1999; the current average funding is $235,000 per year for three years.[15] To support individual-investigator and small-group research grants, the Catalysis Science Program conducts workshops. BES is planning to fund EFRCs, which will probably include increased support for catalysis research.

---

[13]Chemical Sciences Research Programs. Department of Energy Office of Science. http://www.science.doe.gov/bes/chm/Programs/programs.html. Accessed February 2, 2009.

[14]Funds for the Catalysis Science Initiative were made available through the restructuring of existing BES budgets and funds; funds for the HFI were newly appropriated funds that were provided in addition to the existing BES budget.

[15]However, the average grant size for universities was approximately $140,000 per year, whereas the average grant size for national laboratories was approximately $700,000 per year.

3

Overview of the Catalysis Science Program Portfolio

This chapter provides an overview of the U.S. Department of Energy (DOE) Office of Basic Energy Science (BES) Catalysis Science Program's research portfolio in terms of the number and value of research grants and the characteristics of the principal investigators that have been funded by the program since 1987. The review is mainly based on the grant information (institution, project title, and principal investigator) provided by the program manager and on information from the DOE Office of Science Web site.[1] However, demographic data on grantees were obtained from the online version of the *American Chemical Society Directory of Graduate Research* and other Web sites.

BES provided the committee with lists of grantees and project titles for the fiscal years (FYs) 1987 to 2007. In addition, BES designated each grant for FY 1999 to FY 2007 as belonging to one of eight categories: biorelated catalysis, catalysis science, homogeneous catalysis, heterogeneous catalysis, nanoscience, surface science, theory, and other initiatives. However, the committee chose to analyze the portfolio according to the two main kinds of catalysis: heterogeneous (including catalysis science, nanoscience, surface science, theory, and other initiatives) and homogeneous catalysis (including biorelated catalysis).

> **Heterogeneous catalysis.** Grants to study multiphase catalytic reactions, which may contain heterogeneized molecular complexes, with a focus on the role played by the support to which the catalyst is attached. The grants also include those related to surface science, nanoscience, and special funding initiatives, which will be discussed in more detail later in this chapter.

---

[1]Office Of Basic Energy Sciences. U.S. Department of Energy. Online. Available at *http://www.sc.doe.gov/bes/*. Accessed February 5, 2009.

Homogeneous catalysis. Grants to study single-phase reactions involving well-defined molecular catalysts. The research may include heterogeneized molecular catalysts and focus on ligand design. The grants will also be used to study biorelated topics, including biocatalysis, enzyme catalysis, biomimetics, and conversion of biofeedstocks with synthetic catalysts or biological catalysts.

RESEARCH GRANTS

For FY 1999 to FY 2007, in terms of funding and the numbers of grants, heterogeneous catalysis and homogeneous catalysis comprised an average of 70 percent and 30 percent of the Catalysis Science Program portfolio, respectively (Table 3-1). The dollar amount of the grants has been split evenly between national laboratories and universities; however, national laboratories received approximately 17 percent of the grants. This disparity relates to differences between the funding structures of the two types of institutions: national laboratories tend to have greater salary expenditures and higher overhead costs than universities.

BES staff designated the FY 2007 projects (140 projects and 276 principal investigators) by more specific subject of research, as shown in Table 3-2.

BES staff also designated the FY 2007 projects by type of catalyst as follows:

- Inorganic, organic complex catalysts (28 percent)
- Interfacial, porous, hybrid catalysts (35 percent)
- Nanostructured, supramolecular catalysts (27 percent)
- Bioinspired catalysts (3 percent)
- Theory, modeling, simulation relevant to catalysis (7 percent)

Industrial Relevance

The committee members solicited input from their industry contacts (see Appendix D) about the most important breakthroughs in catalysis. Each committee member asked 5 to 10 people in industry who are leaders in catalysis research the question: What has been the most important breakthrough in catalysis within last 20 years? The 32 responses that related to catalysis fell into the following categories, which are roughly aligned with the distribution of the FY 2007 grants mentioned above:

- New materials (e.g., zeolites; nano, micro, meso, macro-porous materials; mixed metal oxides)

- Theory (e.g., density functional theory) and mechanistic understandings
- New methods (analytical, e.g., XAFS; operando)
- New oxidation processes and catalysts (e.g., Au, Pd/Au)
- Environmental catalysis (catalytic converters, emission control catalyst, lean NO_x, low sulfur).

On the basis of the information provided above, the Catalysis Science Program portfolio appears to have a good distribution overall in its funding of basic research in the broad categories of homogeneous catalysis and heterogeneous catalysis. During the time period studied, BES staff did a good job of maintaining the balance between experienced and new researchers within the portfolio by providing stable funding for established researchers while bringing new researchers and topics into the program. However, there are variations in the quality and relevance of the research in the portfolio, which will be discussed in detail in Chapter 5.

TABLE 3-1 Funding and Number of Catalysis Science Program Grants for Research in Homogeneous and Heterogeneous Catalysis, FYs 1999–2001, 2002–2004, 2005–2007

		Funding (millions of dollars)	Percentage of Total Funding	No. Grants
Heterogeneous catalysis				
	1999–2001	48	67%	88
	2002–2004	67	74%	102
	2005–2007	79	73%	137
Homogeneous catalysis				
	1999–2001	24	33%	50
	2002–2004	24	26%	39
	2005–2007	$30	27%	62

NOTE: Grants for research in heterogeneous catalysis (multiphase reactions catalyzed by solid-state catalysts) include nanoscience, surface science, theory, and other initiatives. Grants for research in homogeneous catalysis (single-phase reactions catalyzed by molecular catalysts) include biocatalysis. Data include individual, small-group, and conference grants.
SOURCE: U.S. Department of Energy, Office of Basic Energy Sciences, Catalysis Science Program.

TABLE 3-2 Project Distribution by Subject of Research, FY 2007

Subject	No. Grants
C-H activation	34
Surface properties	31
Inorganic synthesis, inorganic single sites	22
Nanostructure synthesis, properties	21
Oxidation catalysis	20
Ab initio, multiscalar theory	19
Polymerization	12
Electroactivation	10
Surface design	9
X-Y activation	7
H_2 production	7
Ultrafast chemical imaging facilities	5
CHO activation	5
Photocatalysis	5
C-X activation	4
Hybrid sites, cascade reactions	4
Biomass reactions	4
CO, small-molecule hydrogenation	4
Confined complexes	3
Chiral, stereoselective activation	3
Hydrogen storage	2
Aromatic reactions	2
Enzyme chemistry	2
Hydrotreating	2

NOTES: Some grants are related to more than one subject.
SOURCE: Office of Basic Energy Sciences Catalysis Science Program staff.

PRINCIPAL INVESTIGATORS

This section describes the characteristics of the principal investigators who received funding from the Catalysis Science Program during the past eight years (FY 1999 to FY 2007). Table 3-3 shows the distribution of university-affiliated principal investigators according to academic title, and Table 3-4 shows the distribution according to years since receiving a Ph.D.[2] The typical principal investigator was a full professor who had received a Ph.D. approximately 20 years earlier. Twenty-four researchers (approximately 20–30 percent of all grantees) had been funded by the program since 1987. The program also funded a sizable number of junior faculty members; each year, an average of 20 percent of the principal investigators were assistant or associate professors, and 14 percent had received their Ph.D.'s less than 10 years before receiving program funding.

TABLE 3-3 Catalysis Science Program University-Affiliated Principal Investigators by Academic Title, FYs 1999–2007

	1999	2001	2003	2005	2007
Total	87	93	102	122	113
Assistant professor	8	7	9	12	5
Associate professor	8	10	10	15	19
Professor	63	67	70	82	77
Other[a]	8	9	13	13	12

[a]Chair, distinguished professor, adjunct professor, principal scientist, or emeritus professor.
NOTES: Includes national laboratory personnel who are affiliated with a university. Duplicates and grants for conferences were removed from the original BES-supplied lists.
SOURCE: DGRweb 2007. American Chemical Society Directory of Graduate Research. Online. Available at *http://dgr.rints.com*. Accessed March 2008.

TABLE 3-4 Catalysis Science Program University-Affiliated Principal Investigators by Years Since Receiving a Ph.D., FYs 1999–2007

	1999	2001	2003	2005	2007
Total	87	93	102	122	113
Years Since Ph.D.					
≤ 10	15	9	13	18	16
11–20	26	31	22	28	31
21–30	25	27	41	42	37
31–40	20	22	20	26	21
≥ 41	1	4	6	8	8

SOURCE: DGRweb 2007. American Chemical Society Directory of Graduate Research. Online. Available at *http://dgr.rints.com*. Accessed March 2008.

---

[2]The academic title and the year in which the Ph.D. was received were found by using DGRweb. Online version of the *American Chemical Society Directory of Graduate Research*. Online. Available at *http://dgr.rints.com*. Accessed March 2008.

The impact of the Catalysis Science Program on fundamental catalysis science can be measured in part by the reputation of its funded researchers among their peers in the United States and abroad and by the quantity and quality of their journal publications.

Journal Citations

Well-established and respected researchers are typically among the most-cited journal article authors, which appears to be the case for many of the principal investigators who have been funded by the program for a long time. According to the most recent update of the h-index[3] for chemistry (Table 3-5), 18 principal investigators who are currently funded by the program are on the h-index list. More than half of these 18 investigators have received funding from the program for 20 years or more. These scientists are among the best-known chemists in the world and include Nobel laureates. However, the h-index is a measure of past contributions, not current contributions.

U.S. researchers remain among the top performers in terms of number of journal citations, but U.S. representation in catalysis publications has been decreasing. In two of the leading catalysis journals, *Journal of Catalysis* and *Applied Catalysis* (Table 3-6), the number of U.S. papers increased from 1990–1994 to 2000–2006, but the corresponding percentage of the total decreased.[4] At the same time, the European Union share remained stable, and the Asian share increased. Among the 50 papers cited most frequently in these two journals, the number of U.S.-originated papers declined from 27 to 12. In the same report, it was found that this general trend of decline in the share of U.S.-based participation in catalysis was also observed in an analysis of U.S. patents that originated in the United States.

---

[3]In 2005, Jorge Hirsch, of the University of California, San Diego, devised an algorithm, known as the h-index, that calculates the impact of a scientist's work. *Chemistry World* compiles and produces an h-index list of living chemists. A scientist's h-index is equal to the highest number of papers that the person has published that has accrued at least that number of citations. For example, in Table 3-5, G. M. Whitesides has the highest h-index (as of March 28, 2008), 140. That means that he has published 140 papers that have each been cited at least 140 times.

[4]National Research Council. 2007. *International Benchmarking of U.S. Chemical Engineering Research Competitiveness.* Washington, DC: National Academies Press.

TABLE 3-5 H-index ranking of select prominent chemists with relevance to the Catalysis Science Program, according to 2008 H-Index, H-Index Rank, and Field of Chemistry (other prominent chemists listed for comparison)

Rank	Name	H-Index	Field
1	G. M. Whitesides	140	Organic
16	G. Ertl[a]	97	Physical
16	T. Marks[b,c]	97	Inorganic
31	G. Somorjai[b]	92	Physical
36	R. H. Grubbs[b]	90	Inorganic
44	R. R. Schrock[a,b]	87	Inorganic
55	G. A. Olah[a]	83	Organic
72	J. T. Yates[b]	77	Physical
82	L. Que[b]	75	Biological
90	K. Morokuma	74	Theoretical
144	J. E. Bercaw[b,c]	69	Inorganic
152	A. T. Bell[b,c]	68	Physical
152	R. J. Madix[c,d]	68	Physical
167	R. G. Bergman[b,c]	67	Inorganic
188	J. C. Crabtree[b,c]	65	Inorganic
217	D. W. Goodman[b,c]	63	Physical
217	K. N. Raymond[b]	63	Inorganic
232	H. Kroto[a]	62	Physical
253	W. J. Evans[b]	61	Inorganic
272	C. P. Casey[b,c]	60	Inorganic
286	W. Kohn[a]	59	Theoretical
306	D. A. Dixon[b]	58	Theoretical
366	M. H. Chisholm[b,c]	55	Inorganic
366	R. M. Crooks[b]	55	Analytical
435	R. F. Curl[a]	52	Physical
435	L. A. Curtiss[b]	52	Theoretical
435	T. D. Tilley[b]	52	Inorganic
463	T. B. Rauchfuss[b]	51	Inorganic

[a]Nobel laureate.
[b]Receives research funding from the Catalysis Science Program.
[c]Funded by the Catalysis Science Program since 1987.
[d]Emeritus professor, previously funded by BES.
NOTES: List of living chemists compiled by Henry Schaefer, with colleague Amy Peterson, both of the University of Georgia, who assessed h-indexes of approximately 2,000 chemists. The list includes 560 names of those with an h-index greater than 50.
SOURCE: Schaefer, Henry. 2008. H-index ranking of living chemists. Online. Chemistry World. Available at *http://www.rsc.org/chemistryworld/News/2007/April/23040701.asp*. Accessed October 22, 2008.

TABLE 3-6 Papers by U.S. Authors in *Journal of Catalysis* and *Applied Catalysis*

	1990–1994 No. Papers	1995–1999 No. Papers	2000–2006 No. Papers
Total	2,255	3,932	6,859
U.S. Papers	747 (33%)	891 (23%)	1,047 (15%)

SOURCE: National Research Council. 2007. *International Benchmarking of U.S. Chemical Engineering Research Competitiveness.* Washington, DC: National Academies Press.

International Reputation

The Catalysis Science Program funds the most-recognized names in catalysis research, especially in heterogeneous catalysis. This assessment is based on a reputation survey that was conducted for an earlier study to determine the international standing of U.S. researchers in chemistry and chemical engineering.[5] Surveys were carried out separately for heterogeneous and homogeneous catalysis. A new analysis of these surveys that specifically addresses the reputation of Catalysis Science Program grantees is summarized below.

Heterogeneous Catalysis

A total of 331 researcher names, including duplicates, were provided: 140 (42 percent) from the United States and 191 from elsewhere. That 42 percent of the researchers are from the United States indicates that the United States is a leader in heterogeneous catalysis research. Twenty-two names appear on five or more organizer lists, and they are shown in alphabetical order in Table 3-7. Of these, nine are from the United States, and all nine are funded by the Catalysis Science Program (and most with long-term funding). These data imply that the program has a significant impact on the international standing of U.S. research in catalysis and is essential to the success of a heterogeneous catalysis researcher. The dependence of heterogeneous catalysis basic research on the program underlines the crucial role of the program and places substantial responsibility on the program to ensure that the best researchers are identified and supported.

---

[5] See discussion of virtual congresses in National Research Council, 2007, *The Future of U.S. Chemistry Research: Benchmarks and Challenges,* Washington, DC: National Academies Press.

TABLE 3-7 Most-Recognized Heterogeneous Catalysis Virtual Congress Speakers (in alphabetical order)

Name	Affiliation	Country
Baiker, Alfons	ETH Zurich	Switzerland
Barteau, Mark[a]	University of Delaware	United States
Bell, Alexis[a]	University of California, Berkeley	United States
Corma, Avelino	Universidad Politecnica de Valencia	Spain
Dumesic, James[a]	University of Wisconsin, Madison	United States
Ertl, Gerhard	Fritz-Haber	Germany
Freund, Hans-Joachim	Fritz-Haber	Germany
Gates, Bruce[a]	University of California, Davis	United States
Goodman, D. Wayne[a]	Texas A&M University	United States
Hutchings, Graham	University of Wales	UK
Iglesia, Enrique[a]	University of California, Berkeley	United States
Iwasawa, Yasuhiro	University of Tokyo	Japan
Jacobs, P.	Leuven University	Belgium
Lercher, Johannes	Technical University of Munich	Germany
Li, Can	Dalian Institute of Chemical Physics	China
Neurock, Matthew[a]	University of Virginia	United States
Nørskov, Jens	Denmark Technical University	Denmark
Prins, Roel	ETH Zurich	Switzerland
Schmidt, Lanny D.[a]	University of Minnesota	United States
Somorjai, Gabor A.[a]	University of California, Berkeley	United States
Topsøe, Henrik	Haldor Topsøe	Denmark
van Santen, Rutger	Eindhoven University of Technology	Netherlands

[a]Receives research funding from the Catalysis Science Program.
SOURCE: A summarized version of this information appears in National Research Council, 2007, *The Future of U.S. Chemistry Research: Benchmarks and Challenges,* Washington, DC: National Academies Press.

Homogeneous Catalysis

A total of 148 researcher names, including duplicates, were provided for this virtual congress: 77 (52 percent) from the United States and 71 from elsewhere. That 52 percent of researchers are from the United States indicates

TABLE 3-8 Most-Recognized Homogeneous Catalysis Virtual Congress Speakers (in alphabetical order)

Name	Affiliation	Country
Bercaw, John E.[a]	California Institute of Technology	United States
Bergman, Robert G.[a]	University of California, Berkeley	United States
Bianchini, Claudio	Florence	Italy
Brookhart, Maurice	University of North Carolina, Chapel Hill	United States
Buchwald, S.	Massachusetts Institute of Technology	United States
Casey, Charles P.[a]	University of Wisconsin	United States
Crabtree, Roberta	Yale University	United States
Erker, G.	Munster University	Germany
Ferringa, Ben	Groningen	Netherlands
Fu, Gregory	Massachusetts Institute of Technology	United States
Fürstner, Alois	Mülheim	Germany
Gibson, Vernon	London	UK
Grubbs, Robert H.[a]	California Institute of Technology	United States
Hartwig, John A.[a]	University of Illinois	United States
Jacobsen, Erik	Harvard University	United States
Jordan, Richard F.[a]	University of Chicago	United States
Marks, Tobina	Northwestern	United States
Milstein, David	Weizmann Institute	Israel
Noyori, Ryoji	Nagoya	Japan
Okuda, Jun	RWTH Aachen	Germany
Oro, Luis	Zaragoza	Spain
Reetz, Manfred T.	Mülheim	Germany
Schrock, Richard R.[a]	Massachusetts Institute of Technology	United States
Togni, Antonio	ETH Zürich	Switzerland
Trost, Barry M.	Stanford University	United States
Van Koten, G.	Utrecht University	Netherlands
Ziegler, Tom	University of Calgary	Canada

[a] Receives research funding from the Catalysis Science Program.
SOURCE: A summarized version of this information appeared in National Research Council, 2007, *The Future of U.S. Chemistry Research: Benchmarks and Challenges*, Washington, DC: National Academies Press.

that the United States is also a leader in homogeneous catalysis research. Twenty-seven names appear on two or more organizer lists, and they are shown in alphabetical order in Table 3-8. The impact of the Catalysis Science Program on the international standing of U.S. research in homogeneous catalysis is not as evident as in the case of heterogeneous catalysis. Although the percentage of U.S. scientists in the list is higher than the percentage of heterogeneous catalysis researchers, only 8 of the 14 U.S. researchers on the list are currently funded by the program. This is probably due to the fact that multiple sources of funding for homogeneous catalysis research exist in the United States, particularly at the National Institutes of Health. However, several BES-funded principal investigators on the list (such as Bercaw, Bergman, Marks, and Schrock) have made some of the most important contributions to the fundamental understanding of catalysis and to addressing energy needs, which will be discussed in greater detail in Chapters 5 and 6.

CONCLUSIONS

The United States' strong international reputation in catalysis research is built on the records of many established, principal investigators who receive funding from the Catalysis Science Program. However, trends in journal publications indicate that the United States is losing its competitive edge in catalysis, because of increased funding and research efforts in other countries.[6] There is some indication that a gap in contributions from midcareer researchers, especially in heterogeneous catalysis (owing to the downturn in the 1980s' job market), has affected the international standing of U.S. research.

Overall, the Catalysis Science Program portfolio (research grants and principal investigators) is appropriately distributed in its funding of basic research in the broad fields of homogeneous and heterogeneous catalysis. During the time period studied, BES staff did a good job of maintaining a balance within the portfolio and of providing stable funding for established researchers while bringing new researchers and topics into the program. However, there are variations in the quality and relevance of the research in the BES portfolio, which will be discussed in more detail in Chapter 5.

---

[6] National Research Council, 2007, *The Future of U.S. Chemistry Research: Benchmarks and Challenges,* Washington, DC: National Academies Press.

4

Key Influences on the Development of the Office of Basic Energy Sciences Catalysis Science Program Portfolio

The portfolio of the Catalysis Science Program in the Department of Energy (DOE) Office of Basic Energy Sciences (BES) has changed due to a variety of factors. The portfolio is shaped largely by the program manager, with input from the scientific community and the broader public. According to BES, program priorities are established on the basis of

- International trends
- National priorities and laws
- Workshops and professional meetings
- Research solicitations
- Contractor meetings
- BES Advisory Committee (BESAC)

This chapter will describe and evaluate these influences, with the exception of international trends and national priorities and laws.

WORKSHOPS AND PROFESSIONAL MEETINGS

BES staff uses workshops and meetings to bring members of the research community together and to communicate the latest results and directions of its research programs in the chemical sciences, geosciences, biosciences, materials sciences and engineering, and scientific user facilities. BES states that the workshops aid in planning program activities and formulating budget requests. They are also important in integrating efforts among the diverse research communities spread across its research programs and in helping program management to identify new trends and opportunities.

Recently, BES has sponsored two kinds of workshops: those organized by BESAC and those of the Basic Research Needs (BRN) workshop series, which are organized by program managers in the various BES divisions (see Box 4-1). Individual research programs in BES have not typically held separate workshops. Of most relevance to the Catalysis Science Program is the 10th workshop in the BRN series, which focused on catalysis for energy and was held during the summer 2007.

According to BES staff, the chairs for all of the BRN workshops were chosen by the associate director of BES, who considered recommendations by the program and division managers. The primary selection criterion was that the chairs have substantial relevant expertise and seniority so that they would have the confidence of the university and the national laboratory communities. The chairs were charged with identifying panel chairs (subject to BES concurrence), who were typically pairs of university, industry, or national laboratory researchers.

The workshop session topics were selected as a result of discussions between BES and the workshop chairs. The session chairs, in turn, defined the scope and constituency of their panels with input from BES and the workshop chairs. The chairs were encouraged to consider a balanced mix of persons from academe, national laboratories, and industry and were not restricted as to expertise or prior funding from DOE. The total number of participants was limited to avoid unfocused sessions and to encourage maximum interaction among the panelists.

The affiliations of the participants in the BRN workshop on catalysis for energy are summarized in Table 4-1. Approximately 32 percent of the 127 participants were principal investigators funded by the Catalysis Science Program, and 56 percent were principal investigators or others who were not. A number of those outside the Catalysis Science Program had received grants from other BES programs (such as the Solar Photochemistry Program).

BESAC also has held workshops over the years. In 2002, BESAC held a workshop highly relevant to the Catalysis Science Program titled "Opportunities for Catalysis in the 21st Century," which will be discussed later in this chapter.

The Catalysis Science Program also provides support for individual meetings and workshops, such as Gordon research conferences, primarily to support the participation of graduate students and postdoctoral scholars. Recent examples of catalysis-related Gordon conferences supported by the program are those on Catalysis (2002, 2004, 2006, and 2008), Organometallic Chemistry (2002 and 2003), Chemical Reactions at Surfaces (2003, 2005, and 2007), Inorganic Chemistry/Reaction Mechanisms (2004 and 2007), and Electrochemistry (2006 and 2007). The program also has supported other conferences and symposia (for example, the American Chemical Society and the North American Catalysis Society).

> **BOX 4-1**
> **The BES Basic Research Needs Workshop Series**
>
> The 12 reports listed below constitute the integrated output of a set of workshops that were designed by BES to guide strategic planning for the next decade or longer. The series began in 2002 with the workshop titled "Basic Research Needs to Assure a Secure Energy Future," which mapped in general terms the fields of science in which basic research could affect the nation's "decades to century" energy needs. During the subsequent five years, the series of workshops addressed 10 fields in which "use-inspired" basic research (i.e., motivated by potential future applications) might lead to transformative approaches to energy production and use. These fields focused on specific energy-production sectors (hydrogen, solar, and nuclear), uses (solid-state lighting, superconductivity, energy storage, and transportation), and cross-cutting subjects (catalysis, geoscience issues related to energy, and materials under extreme conditions). The final workshop in the BRN series, *Directing Matter and Energy: Five Challenges for Science and the Imagination*, formulated five "grand challenges" that underpin the BES approach.
>
> - How do we control material processes at the level of electrons?
> - How do we design and perfect atom- and energy- efficient synthesis of revolutionary new forms of matter with tailored properties?
> - How do remarkable properties of matter emerge from complex correlations of the atomic or electronic constituents and how can we control these properties?
> - How can we master energy and information on the nanoscale to create new technologies with capabilities rivaling those of living things?
> - How do we characterize and control matter away—especially very far away—from equilibrium?
>
> The BRN workshop reports in ascending order of publication date are as follows:
>
> - *Basic Research Needs to Assure a Secure Energy Future*
> - *Basic Research Needs for the Hydrogen Economy*
> - *Basic Research Needs for Solar Energy Utilization*
> - *Basic Research Needs for Superconductivity*
> - *Basic Research Needs for Solid-State Lighting*
> - *Basic Research Needs for Advanced Nuclear Energy Systems*
> - *Basic Research Needs for Clean and Efficient Combustion of 21st Century Transportation Fuels*
> - *Basic Research Needs for Geosciences: Facilitating 21st Century Energy Systems*
> - *Basic Research Needs for Electrical Energy Storage*
> - *Basic Research Needs: Catalysis for Energy*
> - *Basic Research Needs for Materials under Extreme Environments*
> - *Directing Matter and Energy: Five Challenges for Science and the Imagination*
>
> NOTE: The text of the BRN workshop reports may be found online at http://www.sc.doe.gov/bes/reports/list.html.

TABLE 4-1 Affiliations of Participants in the BRN Workshop on Catalysis for Energy

Affiliation	No. Participants	% of Total
Catalysis Grantees	41	32%
Heterogeneous	29	
Homogeneous	12	
Others	71	56%
National laboratories	31	
Academic	22	
Industry, other	18	
DOE Staff	15	12%
TOTAL	127	

NOTES: Original analysis based on the list of attendees and affiliations in Appendix 3 of *Basic Research Needs: Catalysis for Energy*
SOURCE: Basic Research Needs: Catalysis for Energy. 2007. U.S. Department of Energy Basic Energy Sciences. Online. Available at *http://www.sc.doe.gov/bes/reports/files/CAT_rpt.pdf*. Accessed June 2008.

All of the core research areas of BES have used the BRN workshops, as appropriate, to guide the evolution of their portfolios. In addition, many targeted solicitations in recent years have referred specifically to BRN workshop reports to describe the scientific subjects to be addressed in submitted proposals. The following are examples of grant solicitations that are relevant to the Catalysis Science Program:

- 2008: Energy Frontier Research Centers, http://www.sc.doe.gov/bes/EFRC.html.
- 2008: Single-Investigator and Small-Group Research, http://www.sc.doe.gov/bes/SISGR.html.
- 2006: Basic Research for Solar Energy Utilization, http://www.sc.doe.gov/grants/FAPN06-15.html
- 2006: Basic Research for the Hydrogen Fuel Initiative, http://www.sc.doe.gov/grants/FAPN06-17.html
- 2005: Basic Research in Chemical Imaging, http://www.sc.doe.gov/grants/closed05.html
- 2004: Basic Research for the Hydrogen Fuel Initiative, http://www.sc.doe.gov/grants/Fr04-20.html
- 2003 : Catalysis Science, http://www.sc.doe.gov/grants/Fr03-16.html
- 2001: Nanoscale Science, Engineering and Technology, http://www.sc.doe.gov/grants/Fr02-02.html

Projects relevant to the Catalysis Science Program and funded through the solicitations are discussed in more detail in the next section.

RESEARCH SOLICITATIONS

Since 1999, three initiatives—the National Nanoscience Initiative (NNI), the Catalysis Science Initiative (CSI), and the Hydrogen Fuel Initiative (HFI)—have substantially influenced the direction of the Catalysis Science Program. These initiatives were discussed in Chapter 2 and are briefly described below in Table 4-2. Their influence on the number of grants and the distribution of grants among national laboratories and universities for FY 1999 to FY 2007 is shown in Figure 4-1. The initiatives have led to an increase in funding or an increase in the number of grants in the program portfolio. The CSI has been particularly effective in bringing in new researchers and in fostering more collaborative models of conducting research. The major new source of funding originated from the HFI.

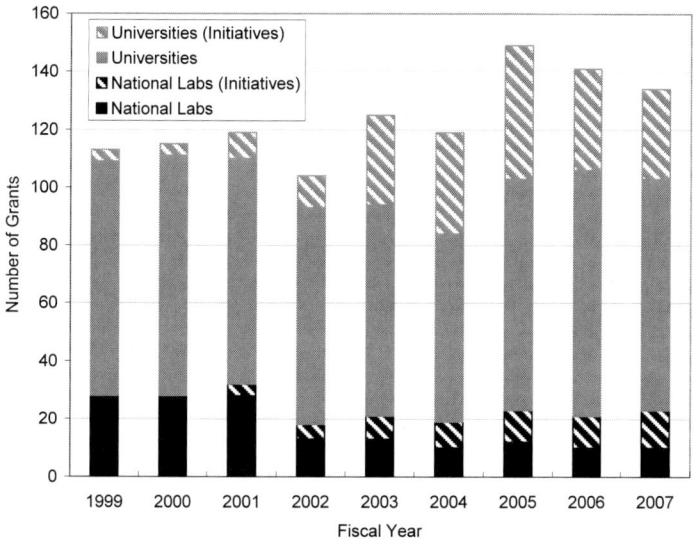

FIGURE 4-1 Influence of initiatives and distribution of Catalysis Science Program grants among national laboratories and universities.
SOURCE: Original analysis based on grant information from Department of Energy, Office of Basic Energy Sciences, Catalysis Science Program.

TABLE 4-2 KEY Initiatives That Have Influenced the BES Portfolio Since FY 2001

Fiscal Year	Initiative	Description
2001	NNI	Grants reorganized and additional funding provided under NNI, which focused on emergent properties at the nanoscale (see http://www.nano.gov for more information); total funding for FYs 2001–2007 was approximately $37.7 million, which included new money.
2003	CSI	Grants funded as a result of a BES solicitation for high-risk, long-term, multi-investigator, multidisciplinary research on the science of catalysis; most grants readapted from single principal investigator to group efforts; total funding for FYs 1999–2007 was approximately $38.1 million from existing BES funds.
2005	HFI	Grants focused on hydrogen production, storage, and use, largely involving electrocatalysis; agency-wide initiative provided new funds and increased the number of grants for the Catalysis Science Program; one-time funding increment of $50 million was allotted to the Chemical Sciences, Geosciences, and Biosciences Division, including a total of approximately $12 million for the Catalysis Science Program.

CONTRACTOR MEETINGS

Unlike the BRN workshops, contractor meetings are organized entirely by individual program managers and are attended primarily by the program's principal investigators. Some recent Catalysis Science Program contractor meetings are as follows:[1]

- *Catalysis Science* (2004)
- *Nanocatalysis Science* (2005)

---

[1] Contractor meeting abstract books are available at http://www.sc.doe.gov/bes/chm/Publications/publications.html.

- *Organometallic, Inorganic, and Bioinspired Chemistry & Catalysis* (2006)
- *Interfacial and Nano Catalysis* (2007)
- *Molecular Catalysis Science* (2008)

The purpose of the meetings is to allow the exchange of information and to encourage cross-fertilization within the program. BES staff says that program managers benefit from a view of the whole program—or major portions of the program—and acquire an impression of overall progress with respect to previous years. The contractor meetings also serve as a forum for the discussion of needs and opportunities in the broader field of catalysis, as well as in the BES catalysis community.

As shown in Figure 4-2, participants in contractor meetings (in contrast with the 2002 BESAC workshop and the 2007 BRN workshop) are mostly principal investigators funded by the Catalysis Science Program. Typically, 60 percent or more of participants in the contractor meetings are catalysis grantees, whereas approximately 30 percent of the participants in the BESAC and BRN workshops are catalysis grantees. Meeting guests include other BES program principal investigators, invited keynote speakers who are not BES principal investigators (including international researchers), and a few invited attendees who are newer researchers at universities or national laboratories not funded by the program. The total number of invitees is restricted by cost and efficiency to about 100 per meeting.

The fields and topics of presentations and discussions at contractor meetings are always chosen to represent both the current program scope and emerging trends. The meetings rotate from year to year (between homogeneous catalysis and heterogeneous catalysis researchers, with the exception of 2004 when the two groups met together), with approximately 60 percent of the program-funded principal investigators attending each year to allow cross-fertilization.

Attendance at the contractor meetings (Table 4-3), particularly the more recent ones, has been appropriately distributed between new grantees (typically junior researchers) and longer-term grantees. At the 2006, 2007, and 2008 contractor meetings and the 2007 BRN workshop, 20 to 25 percent of the grantees attending were new (with funding beginning in the year before, the year of, or the year after the meeting took place). In comparison, less than 20 percent of the catalysis grantees were new to the program at the 2004 and 2005 contractor meetings and the 2002 BESAC workshop.

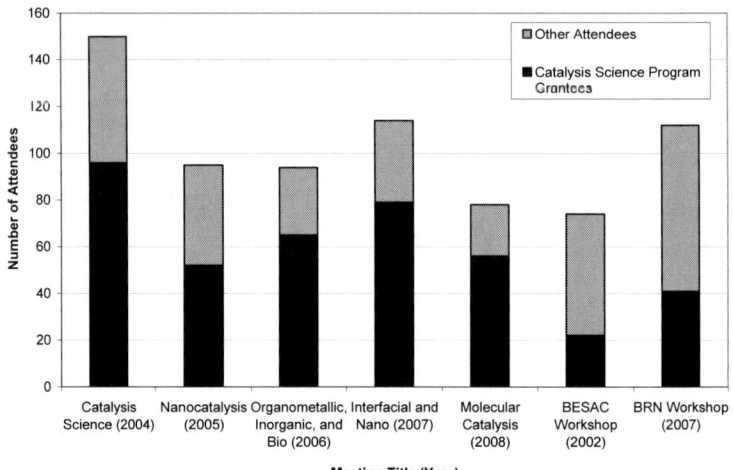

FIGURE 4-2 Distribution of Catalysis Science Program grantees and other attendees at recent contractor meetings (first five bars), the BESAC Workshop, and the BRN Workshop.
SOURCE: Original analysis of lists of meeting attendees.

TABLE 4-3 Attendance at Recent Catalysis Science Program Contractor Meetings and Catalysis-Related Workshops

Meeting Title (Year)	New Grantees[a]	Existing Grantees	Other Attendees	Total Attendees
Catalysis Science (2004)	16	80	54	150
Nanocatalysis Science (2005)	6	46	43	95
Organometallic, Inorganic, and Bioinspired Chemistry & Catalysis (2006)	16	49	29	94
Interfacial and Nano Catalysis (2007)	20	59	35	114
Molecular Catalysis Science (2008)	14	42	22	78
BESAC workshop (2002)	3	19	52	74
BRN workshop (2007)	8	33	71	112

[a]New grantees are those whose funding began in the year before, the year of, or the year after the meeting.

KEY INFLUENCES

The program managers say that they select the chairs (organizers) for the contractor meetings from among the current principal investigators primarily on the basis of their skill as conference organizers. Chairs of the last five Catalysis Science Program contractor meetings are shown in Table 4-4, and they do not seem to follow any particular pattern. The first two meetings in the list (2004 and 2005) were organized by program staff, whereas the last three were organized by a combination of grantees and program staff; the most recent was organized by two newly funded grantees. Chairs of the catalysis-related 2002 BESAC and 2007 BRN workshops are also shown.

According to BES program managers, the major role of a chair is to help with the identification and contacting of keynote speakers and other invited participants. For the latter purpose, they are instructed to obtain external input and advice from the science community, typically department chairs at universities and national laboratories, who are aware of the interests of the newer researchers in their departments.

OFFICE OF BASIC ENERGY SCIENCES ADVISORY COMMITTEE

BESAC was established on September 4, 1986, to provide independent advice to DOE on the complex scientific and technical issues that arise in the planning, management, and implementation of its BES program. BESAC's responsibilities include advising on establishing priorities for research and facilities, determining proper program balance among disciplines, and identifying opportunities for interlaboratory collaboration, program integration, and industrial participation.

One way that BESAC has recently had influence on the catalysis portfolio is through a workshop held in 2002 titled "Opportunities for Catalysis in the 21st Century." According to the summary of the workshop,

> the impetus for the workshop grew out of a confluence of factors: the continuing importance of catalysis to the Nation's productivity and security, particularly in the production and consumption of energy and the associated environmental consequences, and the emergence of new research tools and concepts associated with nanoscience that can revolutionize the design and use of catalysts in the search for optimal control of chemical transformations.

The report concludes as follows:

> [I]n setting its research agenda for the 21st century, the Federal sector must identify and support areas of research that will provide the foundations for the evolution of our current scien-

tific infrastructure into a form that will meet the emerging needs of this new century. That is, the challenges represented on the Federal research agenda for this new century must be sufficiently grand to drive our infrastructure in directions that will meet the uncertainties of the future. One such grand challenge is the development of an understanding, at the molecular level, that will allow us to manipulate, to predict, and ultimately to control chemical reactivity. Catalysis is one of the sciences at the heart of this challenge.

TABLE 4-4 Chairs of Recent Catalysis Science Program Contractor Meetings and Catalysis-Related Workshops

Meeting	Chairs
Contractor Meetings	
Catalysis Science (2004)	John Gordon and Raul Miranda, BES Catalysis Science Program
Nanocatalysis Science (2005)	John Gordon and Raul Miranda, BES Catalysis Science Program
Organometallic, Inorganic, and Bioinspired Chemistry & Catalysis (2006)	Morris Bullock, Brookhaven National Laboratory; Raul Miranda, BES Catalysis Science Program
Interfacial and Nano Catalysis (2007)	John M. White, Pacific Northwest National Laboratory; Charles T. Campbell, University of Washington; Michael J. Chen, BES Catalysis Science Program
Molecular Catalysis Science (2008)	Andreja Bakac and Aaron Sadow, Ames Laboratory and Iowa State University; Michael Chen and Raul Miranda, BES Catalysis Science Program
BES Workshops	
BESAC *Opportunities for Catalysis in the 21st Century* (2002)	J. M. White, University of Texas; John Bercaw, California Institute of Technology
BRN *Catalysis for Energy* (2007)	Alexis T. Bell, University of California, Berkeley; Bruce C. Gates, University of California, Davis; Douglas Ray, Pacific Northwest National Laboratory

SOURCE: Chemical Sciences, Geosciences, and Biosciences Division contractor meeting books:
http://www.sc.doe.gov/bes/chm/Publications/publications.html; BESAC report: http://www.sc.doe.gov/bes/BESAC/reports.html; BRN report: http://www.sc.doe.gov/bes/reports/list.html.

KEY INFLUENCES

BESAC influences individual funding programs through its oversight of the Committee of Visitors (COV). In 1999, the director of the Office of Science charged BESAC with establishing a COV to monitor active projects and programs and to provide regular assessments of the processes used to solicit, review, recommend, and document proposal actions. COVs were assembled to evaluate the BES Chemical Sciences, Geosciences, and Biosciences Division (CSGB) in 2002, 2005, and 2008. The 2005 COV provided the most detailed evaluation of the Catalysis Science Program.

The 2005 COV found CSGB as a whole to be well managed and in excellent shape, with clear evidence that the processes for proposal solicitation, review, and action were working well and that the quality of science, the depth and breadth of portfolio elements, and the national and international standing of the elements were very good to excellent in all nine programs reviewed. However, the COV identified two issues that required substantial attention by CSGB and BES management. One was the continuing lack of an integrated Office of Science database on processes for proposal review, tracking, decision and documentation, and monitoring (of funded proposals) and the lack of standardized database software that would allow rapid and efficient searches for information on principal investigators, reviewers, proposal actions, and principal investigators' productivity. The second was the need for the Office of Science to implement ways to track gender and racial diversity among principal investigators in universities and DOE laboratories and among reviewers. The COV noted that continuing inattention to diversity would potentially have a long-term adverse effect on workforce development.

The 2005 COV provided detailed findings and recommendations about the efficacy and quality of the Catalysis Science Program's processes and the effects of the award process on the portfolio (see Appendix E). The results of the 2005 COV assessment of the Catalysis Science Program can be summarized in four categories: program quality, the program manager, contractor meetings, and funding mechanisms. As pointed out in the COV report, the overall quality of the Catalysis Science Program is high (see Findings and Conclusions in Chapter 6 of this report), and the program has no doubt been favorably influenced by having the same program manager for the past eight years. The contractor meetings are important for monitoring the program and influencing the makeup of the funding portfolio. The Catalysis Science Program continues to struggle to find the right balance of funding between single and groups of investigators.

CONCLUSIONS

The Catalysis Science Program is responding to external factors and using internal mechanisms to shape its research portfolio. In recent years, the program has been substantially influenced by special funding initiatives (for example, the Hydrogen Fuel Initiative). The program manager has done a good job of applying the initiatives in a way that provides continued support for long-term grantees (and basic science) while bringing new researchers and new research approaches into the portfolio. The program also has been successful in using advice from BESAC (derived through workshops and COVs), contractor meetings, and professional meetings to shape its portfolio and influence funding decisions in a more "bottom-up" approach.

The committee recommends that the Catalysis Science Program continue to broaden participation in its contractor meetings and other activities. Non-DOE sponsored workshop organizers and grantees funded by other BSE programs should be invited to attend the Catalysis Science Program activities to provide a more diverse influence on the portfolio. This is particularly important in the development of research directions that will have a long-term impact on the program.

5

Analysis of the Catalysis Science Program Portfolio

This chapter evaluates the research portfolio of the U.S. Department of Energy (DOE) Office of Basic Energy Sciences (BES) Catalysis Science Program in detail.

METRICS FOR EVALUATING THE CATALYSIS SCIENCE PROGRAM PORTFOLIO

In the assessments below, grants are evaluated in terms of the characteristics of the portfolio, their impacts on fundamental science, and their contributions to meeting near-term and long-term national energy goals. The second and third metrics are explained below.

Fundamental Advances in Science

Defining *fundamental science* is challenging. "I know it when I see it" is an adage to which many scientists would agree. However, the committee has been asked to judge whether the Catalysis Science Program has "advanced fundamental science in catalysis," so we must have some generally shared definition of *fundamental science in catalysis*. For the purposes of this study, we will define *fundamental science in catalysis* as a general understanding of or insight into a catalysis system or material that is fundamental enough to be applied to more than one catalyst. Examples might be the development of quantitative models of a class of reactions (such as hydrocarbon oxidation) using a class of catalysts (such as noble metals), the synthesis of a new class of materials (such as zeolites), and the understanding of reaction or surface mechanisms of a class of catalysts (such as transition metal oxides).

Contributions to the Nation's Energy Goals

The effectiveness of the Catalysis Science Program can also be judged by measuring its contribution or likely contribution to meeting near-term and long-term national energy goals. Near-term and long-term national energy goals have been neither clearly stated nor static during the past 20 years. However, for the purposes of this study, we will consider energy goals to be any goals related to reducing the amount of energy we need (efficiency) or reducing our need to import oil immediately or over the long term.

As described in Chapter 3, the Catalysis Science Program portfolio is distributed between the two main categories of catalysis: heterogeneous and homogenous, each of which will be assessed separately below. The committee has made this distinction for convenience, based on the traditional division in catalysis. However, researchers are increasingly crossing the traditional barriers between heterogeneous and heterogeneous catalysis, blurring the distinction between the two (see the discussion of Contractor Meetings in Chapter 4), which the committee definitely views as a positive development. The names of principal investigators are provided in the assessments where appropriate, along with references to contractor meeting abstract books or published journal articles. Lists of all of the principal investigators who received funding during the fiscal years (FYs) 1999 to 2007 and FY 2008 are provided in Appendix F.

HETEROGENEOUS CATALYSIS

Heterogeneous catalysts are the catalysts most commonly used in industrial processes. Heterogeneous catalysis involves the use of a catalyst (which is typically a solid) that is in a different phase from the reactants (which are typically gases). Heterogeneous catalysts range in composition from solid metals to encapsulated metal nanoparticles in solution.

Virtually every drop of oil is in contact with multiple heterogeneous catalysts during the refining process. Most commodity chemicals are produced by heterogeneously catalyzed processes. Heterogeneous catalysis is also used extensively in environmental processes. Every car now has a heterogeneous catalytic converter in its exhaust system to remove toxic fumes. Heterogeneous catalysis is used to clean flue gases from power plants and to remove toxic gases or odors from industrial production. The use of heterogeneous catalysts to remove sulfur compounds from oil products and to clean flue gases from power plants are the primary reasons for the considerable decrease in the amount of acid rain in recent years.

Why Is Heterogeneous Catalysis Important for Energy?

Heterogeneous catalysis is a key to developing energy-efficient, environmentally friendly processes for the conversion of fossil energy feedstocks (coal, petroleum, natural gas, and tar sands) to usable products, including gasoline and diesel oil. Because heterogeneous catalysis is heavily used in the chemical industry, its efficiency is extremely important. To understand its importance, consider the catalytic production of ammonia, which is used to make fertilizers and is instrumental in the world's food production; this process alone accounts for more than 1 percent of the world's energy consumption.[1]

Heterogeneous catalysis is also essential for new, sustainable energy processes. It is used to convert biomass to usable energy products. It also is a key to new photochemical and electrochemical processes for possible future sustainable production of fuels based on sunlight and for new efficient processes for using such fuels in fuel cells. In many cases, it is the lack of efficient catalysts that prevents new technologies for using chemical energy from being economically feasible today.

Assessment of Heterogeneous Catalysis Portfolio

The committee assessed the grants within the heterogeneous catalysis portfolio according to the subareas described below, which are based on how the grant information was provided by the Catalysis Science Program staff.

> *Traditional heterogeneous:* Individual investigator grants involving heterogeneous catalysis but not associated with any new programs or initiatives.
>
> *Surface science:* Grants focused on the understanding of heterogeneous catalytic surfaces and the advanced application of surface science or the development of new approaches within surface science.
>
> *Nanoscience*: Grants funded under the National Nanotechnology Initiative (NNI; *http://www.nano.gov*) or grants focused on emergent properties at the nanometer scale. The NNI was established in FY 2001 to coordinate federal nanotechnology research and development.
>
> *Catalysis Science*: Grants funded under the Catalysis Science Initiative (CSI), first awarded in 2003 and to multi-investigator, multidisciplinary

---

[1] Worrell, E., D. Phylipsen, D. Einstein, et al. 2000. *Energy Use and Energy Intensity of the U.S. Chemical Industry*. Lawrence Berkeley National Laboratory (LBNL-44314). Online. Available at *http://industrial-energy.lbl.gov/node/86*. Accessed February 2, 2009.

teams. The stated goal of the CSI is to "develop combined experimental and theoretical approaches to enable molecular-level understanding of catalytic reaction mechanisms, ultimately enabling the prediction of catalytic reactivity at multiple time and length scales."

Theory: Grants focused on theory, modeling, or simulation. Grants in other categories may also include theory but not as the focus of the research.

Other initiatives: Grants funded by various DOE cross-cutting programs. During FY 1999 to FY 2001, coal chemistry and noncatalytic reactions were included in this category. More recently, grants have included those under the Hydrogen Fuel Initiative (HFI; established in 2005) and in related fields, such as electrochemistry.

The distribution of grants into the six subareas of the heterogeneous catalysis portfolio for the time period studied is provided in Table 5-1.

Overall, during the past eight years (FY 1999 to FY 2007), there appears to be appropriate retention in research topics with a slight shift in focus to such growing fields as nanoscience and theory. Considering the central role that catalysis plays in traditional and alternative fuel production and the level of funding that is available for the Catalysis Science Program, BES has done an impressive job during the past eight years of providing the nation with fundamental research in heterogeneous catalysis.

TABLE 5-1 Distribution of Grants in the Heterogeneous Catalysis Portfolio by Subarea

	Traditional Heterogeneous	Surface Science	Nanoscience	CSI	Theory	Other, HFI
1999–2001	28	22	13	---	3	16
2002–2004	26	19	20	13(24^a)	6	4
2005–2007	22	21	14	13(23^a)	17	16(14 HFI)

^{a}Number of participating institutions.

Traditional Heterogeneous

Within this subarea, a variety of projects are addressing the fundamental questions in heterogeneous catalysis related to the nature of active sites and structure–activity relationships. The projects can be characterized according to the class of materials being studied, such as metals supported on one or two metal oxides, metal oxides with various promoters, and specific zeolitic materials; and the type of reaction being studied, such as methanol synthesis, olefin oxidation, and so on. These projects are tackling problems of immediate interest to industry because they typically are looking at advanced systems of present catalysts, such as automotive catalysts; or at new catalysts for important (large) processes, such as methanol synthesis. Traditional heterogeneous projects have often evolved over several grant cycles, and the evolution has tended toward more sophisticated characterization techniques (including density functional theory calculations) and a considerably deeper understanding of specific reactions using a particular catalyst. In many cases, this deeper insight has opened the possibility of suggesting new, more active catalysts.

Analysis of the Traditional Heterogeneous Subarea. This subarea has provided a long-term stable funding basis for a number of U.S. researchers that are the dominant figures in the field. Some were principal investigators in the traditional heterogeneous catalysis subarea throughout the period studied. This group includes A. Bell, J. Dumesic, B. Gates, R. Gorte, G. Haller, H. Kung, S. Suib, and I. Wachs (see Appendix Table F-1). A clear strength of the program is that it has provided a stable funding environment for leading researchers in the field. The disadvantage is that the program has at times been perceived as fostering a relatively closed community by providing too few opportunities to new researchers. That perception was aggravated by earlier funding practices in the Catalysis Science Program and by the economic situation in the 1980s.

Only 17 new awards were granted to heterogeneous catalysis researchers during the nine years following 1987. The lack of funding for new researchers through the program coincided with a drop in hiring beginning in 1986 in the petroleum industry, a major employer of catalysis scientists at that time. The lack of opportunity in heterogeneous catalysis research seems to have had a lasting effect on the community, particularly considering the lag time involved in training new Ph.D. researchers. More recently, significant efforts to improve the balance between new and established researchers have been successful. However, because of the opportunity gap that existed during the mid 1980s to late 1990s, a "generation gap" in heterogeneous catalysis researchers still remains.

Impacts on Fundamental Science and Contributions to Energy Goals. Examples of funded research in this subject can be found in the 2007 Catalysis Science

Program contractor meeting summary.[2] They include the work of B. Gates on providing models for understanding heterogeneous catalysis mechanisms, the work of E. Iglesia on the design of single-site catalysts for oxidation reactions, studies by R. Gorte on cerium-based catalysts for automotive catalyst applications, and the work of A. Bell on methanol catalysis and metal oxides (Cu/ZrO_2).

There is an important class of projects that are exploring new synthesis methods, unique reactor systems, unique characterization techniques, or a new set of reactions. Examples include the work of L. Schmidt, who has done pioneering work on short-residence-time reactors; the work of R. J. Davis, who has studied basic and acidic properties of catalysts by using various probes and spectroscopic methods and by using theory to enhance the work;[3] and the work of J. A. Dumesic on liquid-phase reforming of biomass for energy purposes (Box 5-1). The latter work illustrates how the Catalysis Science Program can support research leading to completely new catalytic chemistry and a series of new catalysts.

Surface Science

Over the past four decades, the field of surface science—including reaction, spectroscopic, and imaging studies of single crystals and other model surfaces—has brought the description of reaction mechanisms, intermediates, and active sites in heterogeneous catalysis from schematic depictions to observable structures. It has contributed substantially to the science of heterogeneous catalysis and in recent years has provided much of the critical experimental database with which to benchmark electronic structure calculations. The award of the 2007 Nobel prize in chemistry to Gerhard Ertl of Germany recognized the impact of surface science on the understanding, design, and control of catalytic processes.

---

[2] Basic Energy Sciences. 2007. *Frontiers in Interfacial and Nano Catalysis*. U. S. Department of Energy and Oak Ridge Associated Universities. Online. Available at *http://www.sc.doe.gov/BES/chm/Publications/Contractors%20Meetings/2007_Catalysis.pdf*. Accessed January 13, 2009.

[3] Siporin, S. E., B. C. McClaine, and R. J. Davis., 2003. Adsorption of N_2 and CO_2 on zeolite X exchanged with potassium, barium or lanthanum. *Langmuir* 19:4707-4713; and Li, J., J. Tai, and R. J. Davis. 2006. Hydrocarbon Oxidation and Aldol Condensation Over Basic Zeolite Catalysts. *Catalysis Today* 116:226-233.

> **Box 5-1**
> **Traditional Heterogeneous Catalysis Contribution to Energy Goals**
>
> The conversion of biomass for energy purposes has opened up a field of research that may have a substantial impact on the advancement of science and on progress toward meeting the nation's energy goals. In 2002, J. A. Dumesic and colleagues discovered that aqueous solutions of oxygenated hydrocarbons with a C:O stoichiometry of 1:1 could be converted with high yields over platinum-based catalysts at temperatures near 520 K to gas mixtures of hydrogen and carbon dioxide containing low concentrations of carbon monoxide (such as 50 ppm).[1] That discovery was inspired by initial work in the Dumesic group dealing with the selectivity for cleavage of C-O versus C-C bonds in oxygenated hydrocarbon intermediates on platinum surfaces. The Dumesic group continued its work in aqueous-phase reforming to target the conversion of sugars and polyols to alkanes.[2] Specifically, it targeted the selective cleavage of C-O bonds versus C-C bonds. The conversion of a sugar or polyol to an alkane having the same number of carbon atoms as the original reactant was a major advance in the conversion of biomass resources to liquid transportation fuels. Ensuing investigations explored various routes for achieving C-C coupling reactions between biomass-derived intermediates before the final removal of all remaining hydroxyl groups to form longer-chain liquid alkanes by dehydration combined with hydrogenation. In more recent work, Dumesic's group has designed new catalysts (inspired by results of density-function theory calculations in the literature) to carry out the conversion of glycerol to synthesis gas at temperatures near 540 K. In particular, high reaction rates have been achieved at low temperatures and at pressures near 20 atm with new Pt-Ru and Pt-Re bimetallic alloys.[3]
>
> [1] Cortright, R. D., R. R. Davda, J. A. Dumesic. 2002. Hydrogen from Catalysis Reforming of Biomass-Derived Hydrocarbons in Liquid Water. *Nature* 418:964-967.
> [2] Huber, G. W., R. D. Cortright, J. A. Dumesic, 2004. Renewable Alkanes by Aqueous Phase Reforming of Biomass-Derived Oxygenates. Angew. *Chem. Int. Ed.* 43:1549-1551.
> [3] Soares, R. R., D. A. Simonetti, J. A. Dumesic. 2006. Glycerol as a Source for Fuels and Chemical by Low-Temperature Catalytic Processing. Angew. *Chem. Int. Ed.* 45:3982-3985.

Recently, Dumesic and his potential impact on the biorefinery concept were noted in a *Science* profile,[4] and his work was featured in *Green Chemistry* (Figure 5-1). The University of Wisconsin has obtained several patents based on the group's research.

---

[4] Cho, A. 2007. Profile: James Dumesic: Catalyzing the Emergence of a Practical Biorefinery. *Science* 315:795.

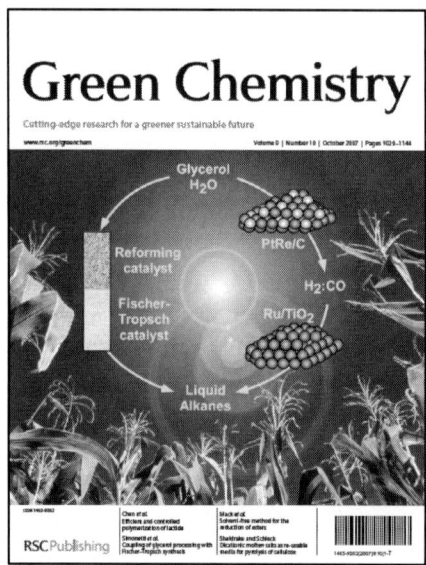

FIGURE 5-1 Integrated process to produce liquid fuels directly from glycerol. Liquid fuels can be produced in single reactor by coupling low-temperature conversion of glycerol to synthesis gas with formation of liquid alkanes by Fischer–Tropsch synthesis.
SOURCE: Simonetti, D. A., J. Rass-Hansen, E. L. Kunkes, R.R. Soares, and J.A. Dumesic. 2007. Coupling of glycerol processing with Fischer–Tropsch synthesis for production of liquid fuels. *Green Chemistry* 9:1073-1083.

Analysis of the Surface Science Subarea. Since its inception, the Catalysis Science Program has supported U.S. leaders in this field, such as G. Somorjai, R. Madix, and W. H. Weinberg (see Appendix Table F-1). It has also supported the growth of the field through a second generation of principal investigators (such as M. Barteau, C. Campbell, C. Friend, and W. Goodman), many of whom were graduate and postdoctoral students of the science's pioneers. Collectively, the principal investigators of the surface science portion of the portfolio have garnered an impressive number of international prizes, National Academies memberships, and awards from the American Chemical Society, the American Institute of Chemical Engineers, and various catalysis societies. However, the mix of principal investigators in the surface science subarea appears to be fairly static and aging; for example, there has been a decrease in the number of those who received their Ph.D.'s 11–20 years before receiving program funding (Table 5-2). As a result, the "generation gap" in the field of catalysis and in the Catalysis Science Program portfolio noted elsewhere in this report is more of a demographic "cliff" in the surface science subarea. However, the reality is encouraging: although there has been some erosion of U.S. competitiveness in surface

TABLE 5-2 Distribution of Surface Science Principal Investigators in the Catalysis Science Program Portfolio by Years Since Receipt of Ph.D.

Years since Ph.D.	Frequency Distribution				
	1999	2001	2003	2005	2007
≤ 10	2	0	1	1	1
11–20	10	9	5	5	5
21–30	3	5	8	8	6
31–40	4	4	3	2	1
≥ 41	0	2	2	4	3
Total	19	20	19	20	16

science relative to Europe and Asia, there has also been a healthy evolution of the field that is well represented in the Catalysis Science Program portfolio, even if not in the static classification of surface science programs within the portfolio.

The percentage of projects classified as primarily surface science in the program portfolio has remained fairly constant over the past decade, but a proper assessment of the field must account for other programs in BES that contain substantial support for surface science. For example, the Condensed Phase and Interfacial Molecular Science Program, the Materials Chemistry Program, and the Solar Photochemistry Program have important surface science components. Collectively, these three programs support nearly the same number of projects that could be classified as surface science aimed at catalysis as are supported by the Catalysis Science Program. Furthermore, the number of subfields encompassed by the program has increased, in part due to the strong pull of new and important applications of surface science research. A substantial percentage of the nanoscience projects in the program's portfolio can be credited to principal investigators in surface science and their students. An aspect of these two developments is an apparent generation gap in the United States that may include much of the broader field of catalysis science.

For example, the work of C. Campbell (see Appendix Table F-1), the current editor of the journal *Surface Science*, is now considered to be part of the nanoscience portfolio. The same is true of theory: a substantial portion of the growth in the theory portfolio, especially in projects that incorporate both experiments and theory, has spun out of surface science groups—both past principal investigators and their students. Thus, looking at a broader portfolio of projects that might be labeled "model systems" and seen as encompassing surface science and nanoscience, including dynamics and theory, we can see an evolution of the research programs and a refreshing of the principal investigator pool with younger researchers. There are programs that integrate surface science, theory, nanoscience, and catalysis in the CSI portfolio. For example, *all* of the current CSI projects have a substantial computation and theory component, and approximately one-half have a surface science component.

Impact on Fundamental Science. As noted above, research and researchers funded by the surface science portion of the portfolio have contributed to the growth of nanoscience and theory. Historically much of heterogeneous catalysis and the research supporting it have been at the nanoscale. However, the increased and broader focus on nanoscience at the national level has changed the "center of gravity" in surface science. During the most recent three-year time period, approximately one-half of the projects focused primarily on surface reaction mechanisms, and the other half focused more on surface structure.
That leaves perhaps fewer than 10 groups in the country addressing surface reaction mechanisms as part of the surface science portfolio, and all of the principal investigators are established full professors. One consequence is that U.S. contributions to *Surface Science* typically make up 10–20 percent of the content per issue, and fewer still are related to studies of reaction mechanisms relevant to catalysis.

Contributions to Meeting Energy Goals. During the past decade, the principal investigators in the surface science subarea have made numerous contributions to the mechanistic and structural understanding of catalytic reactions that continue to advance catalysis of reactions and processes with energy implications. The work has provided the crucial foundation for the grand challenge, "Understanding Mechanisms and Dynamics of Catalyzed Transformations," that was articulated in the recent workshop report, *Basic Research Needs in Catalysis for Energy*. Examples of areas of impact include the topics listed below with the names of the Catalysis Science Program principal investigators provided in parentheses[5].

- Hydrogenation and dehydrogenation (P. Stair, B. Koel, J. Vohs, and F. Zaera)
- Hydrocarbon reforming (G. Somorjai)
- Oxygenate reforming (J. Chen)
- Selective oxidation (M. Barteau, C. Friend, and W. Tysoe)
- Heteroatom removal (C. Friend, F. Ribeiro, G. Somorjai, and J. Vohs)
- Surface photochemistry and catalysis (U. Diebold, P. Stair, J. Yates)

---

[5] Basic Energy Sciences. 2004. *Frontiers in Catalysis*. U.S. Department of Energy and Oak Ridge Associated Universities. Online. Available at
http://www.sc.doe.gov/BES/chm/Publications/Contractors%20Meetings/CatalysisContrMtg2004/AbstractBook/StartHere.htm. Accessed January 13, 2008.

- Structure and dynamics of catalyst surfaces (E. Altman, C. Campbell, C. Friend, T. Madey, R. Madix, J. Vohs, J. Weaver, J. M. White)
- Bimetallic and alloy systems (J. Chen, W. Goodman, B. Koel, J. Vohs)

Nanoscience

The National Nanotechnology Initiative (NNI) began funding catalysis research in 2001. At that time, BES identified two types of catalysis-related nanoscience: one type funded by the NNI and another type funded by BES general funds but identified as focusing on nanoscience, both of which will be addressed in this section.

Analysis of the Nanoscience Subarea. The NNI brought additional funding to the Catalysis Science Program: 10 awards were given—3 to national laboratories and 7 to universities. The principal investigators chosen to receive the additional NNI funding in FY 2001 included well-established catalysis researchers (G. Somorjai, J. M. White, and C. Nickolls), but 3 awards were given to newly appointed professors (see Appendix Table F-1). Of the 10 projects funded in FY 2001, 7 were still being funded in FY 2007, in keeping with the BES trend of providing long-term funding. In FY 2002, 14 awards were given: 9 were continuations to the principal investigators funded by the NNI in FY 2001 (1 of the national laboratory proposals was not refunded), and 5 were new. Of the 5 new projects in FY 2002, 4 were to established researchers, but only 2 of them remained funded in FY 2007. Overall, the influx of NNI funding into the Catalysis Science Program led to funding of four new professors in nanoscience, or approximately one-fourth of the new funding.

The Catalysis Science Program funds nanoscience work that is directly related to catalysis and to work that is more fundamental. Although the fundamental work, such as the metal-organic framework synthesis, may be high risk, it is necessary and valuable if the next breakthrough materials—such as the next "zeolites"—for catalysis are to be discovered.

Impacts on Fundamental Science. Most of the NNI projects focus on synthesis of novel single-site heterogeneous catalysis, nanoparticles as catalysts, or new materials that might lead to a new family of catalysts. New materials are explored through novel synthesis schemes that are used to make porous solids for use as catalysts or by reacting catalytic species with potential catalyst supports. A good example of such work is that of A. Maverick and colleagues (Figure 5-2), who are constructing porous inorganic–organic hybrid molecules that serve as a framework for solids that contain coordinately unsaturated metal centers. Although the materials are not constructed specifically to be catalysts, they illustrate the development of porous new materials that might lead to catalysis breakthroughs. The group has synthesized a two-dimensional honeycomb material by using iron (III) complex Fe(Pyac)3 with Ag as a link to yield a bimetallic crystalline solid.

The nanoscience portfolio contains many projects that involve interesting materials and synthesis schemes (for example, the metal-organic framework materials mentioned above) that could lead to the next generation of catalytic materials. The need to have consistently identical single-site catalysts to promote the most selective catalysts is the basis for several of the nanoscience projects.

Another theme is the synthesis and manipulation of nanoparticles to be catalysts. Researchers at Oak Ridge National Laboratory (Figure 5-3) are studying the effects of the size, synthesis, and other characteristics of supported gold nanoparticles on catalyst activity and selectivity.

Finke and coworkers are studying the growth mechanisms of nanoclusters, how to stabilize transition-metal nanoclusters, and how they can be used as catalysts. Their materials include an acetone hydrogenation catalyst made from $Ir(0)_n$ nanocluster and a $Pt(0)_n$ hydrogenation catalyst.

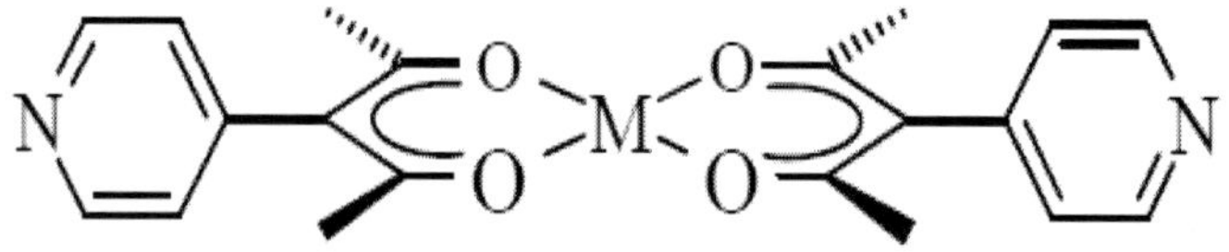

FIGURE 5-2 Rod-shaped building block of M(Pyac)2.
SOURCE: A. W. Maverick. 2006. Inorganic-Organic Molecules and Solids with Nanometer-Sized Pores. In Frontiers in Organometallic, Inorganic, and Bioinspired Chemistry & Catalysis. U.S. Department of Energy and Oak Ridge Associated Universities. Online. Available at
http://www.sc.doe.gov/bes/chm/Publications/Contractors%20Meetings/2006_Catalysis.pdf. Accessed January 13, 2008.

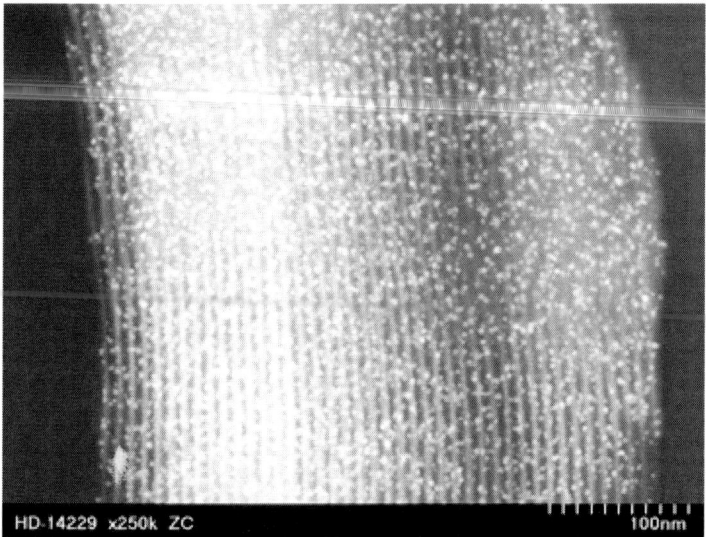

FIGURE 5-3 Immobilization of gold cationic precursor on a negatively charged surface of SBA-15 led to uniform Au particles of 2.9 nm dispersed in 7-nm pores of SiO_2. After activation, catalyst was highly active for CO oxidation.
SOURCE: Overbury, S. H. 2006. Nanocatalysts: Synthesis, Properties, and Mechanisms. *Frontiers in Organometallic, Inorganic, and Bioinspired Chemistry & Catalysis.* U.S. Department of Energy and Oak Ridge Associated Universities. Online. Available at http://www.sc.doe.gov/bes/chm/Publications/Contractors%20Meetings/2006_Catalysis.pdf. Accessed January 13, 2008.

Catalysis Science Initiative

The CSI was launched in 2003 in response to the 2002 BES Advisory Committee (BESAC) workshop report,[6] which identified the grand challenge for catalysis science in the 21st century to be developing an understanding of how to design catalyst structures to control catalytic activity and selectivity. The ability to design catalysts, rather than empirically hunt for the proper catalyst, has long been an agreed-upon goal of the catalysis community. Besides identifying that grand challenge, the report clearly and succinctly documented the need to bring a multidisciplinary approach to the design of catalysts.

Analysis of the Catalysis Science Initiative Subarea. The CSI focuses on collaborative research teams of 3 to 10 principal investigators. The CSI accounted

---

[6] Basic Energy Sciences Advisory Committee. 2002. *Opportunities for Catalysis in the 21st Century*. U.S. Department of Energy. Online. Available at http://www.sc.doe.gov/BES/reports/files/OC_rpt.pdf. Accessed January 30, 2009.

for 18 percent of the program for FY 2002 to FY 2004 and 20 percent for FY 2004 to FY 2007. Eleven teams were funded in response to the initial call for proposals in FY 2003. They included 9 teams for which a university was the lead institution and two for which a national laboratory was the lead institution. There were 59 principal investigators in total in 24 institutions. Six of the initial 11 awards involved cross-institution collaboration. Two additional awards were made in FY 2004: one to a university-led team and one to a national laboratory team.

All of the awards were for a three years, and all of the projects funded in FY 2003 to FY 2004 were required to submit renewal proposals for peer review in FY 2006 to FY 2007. Ten of the 11 FY 2003 projects and 1 of the 2 FY 2004 projects were renewed. The remaining 10 original programs will be required to submit proposals for renewal early in 2009.

The principal investigators of the CSI-funded groups have been both established leaders (M. Barteau, N. Delgass, and F. Zaera) and potential leaders in catalysis (S. Scott and T. Heinz).[7] Substantial effort has been made by group leaders and by BES program directors to reach out to involve new investigators. Of the nine university awards in FY 2003, three were to investigators with previous program funding, and six were to investigators new to the program. Of the additional 34 co-principal investigators listed in the awards, more than one-half appear to have been new to the program. Several of the new investigators have since been successful in obtaining additional support, especially through the Hydrogen Fuel Initiative. Thus, the CSI appears to have been quite successful in attracting and supporting new investigators in catalysis during the past five years.

The 2002 BESAC report, *Opportunities for Catalysis in the 21st Century,* identified the following grand challenge in catalysis: "to understand how to design catalyst structures to control catalytic activity and selectivity." The proposals funded under the CSI clearly reflect the response of researchers in the field and of program management to that challenge. The titles of the 11 awards made in FY 2003 are:

- From First Principles Design to Realization of Bimetallic Catalysts for Enhanced Selectivity
- Electron Transport, Oxygen Activation and Biosynthesis: An Integrative Electrochemical and Computational Approach
- Catalyst Design by Discovery Informatics
- Controlling Structural, Electronic, and Energy Flow Dynamics of Catalytic Processes through Tailored Nanostructures

---

[7] Basic Energy Sciences. 2007. *Frontiers in Interfacial and Nano Catalysis.* U.S. Department of Energy and Oak Ridge Associated Universities. Online. Available at http://www.sc.doe.gov/BES/chm/Publications/Contractors%20Meetings/2007_Catalysis.pdf. Accessed January 13, 2008.

- Basic Principles That Govern the Interaction of Organometallic Catalysts with Supports—the Science of Immobilized Molecular Catalysts
- Principles of Selective O_2-Based Oxidation by Optimal (Binuclear) Catalytic Sites
- Hierarchical Design of Heterogeneous Catalysis for Hydrocarbon Transformations: Structures and Dynamics of the Active Sites
- The Reactivity and Structural Dynamics of Supported Metal Nanoclusters Using Electron Microscopy, In Situ X-Ray Spectroscopy, Electronic Structure Theories, and Molecular Dynamics Simulations
- Molecular-Level Design of Heterogeneous Chiral Catalysts
- Early Transition Metal Oxides as Catalysts: Crossing Scales from Clusters to Single Crystals to Functioning Materials
- Selective and Efficient Catalysis in 3-D Controlled Environments

Collectively, these grants represent the leading aspirations in the field—to design new and improved catalysts from first principles, to integrate homogeneous and heterogeneous catalysts, to understand and control the dynamics of catalytic processes, and to incorporate information and discoveries from biological systems, theory, computation, and nanoscience to advance catalysis science (for example, see Box 5-2). These aspirations are echoed by the grand challenges articulated in the recent (2008) BES reports, *Directing Matter and Energy: Five Challenges for Science and the Imagination* and *Basic Research Needs: Catalysis for Energy*.

Research awards under the CSI have incorporated elements of all of the other portions of the Catalysis Science Program portfolio: homogeneous catalysis, heterogeneous catalysis, surface science, nanoscience, biorelated catalysis, theory, and electrocatalysis. However, as in the portfolio as a whole, integration has tended to be greater in heterogeneous than in homogeneous catalysis. Thus, the programs currently funded under the CSI focus on heterogeneous catalysis, although several involve molecular complexes, clusters, and catalysts, both in solution and on supports.

All 11 projects currently funded under the CSI are less than five years old. Representing approximately 20 percent of the current portfolio, they have already produced scientific successes such as those noted above and in demonstrating the integration of the disciplines across catalysis science and in attracting and supporting investigators new to the field. This record suggests that the initiative has been a valuable addition to the Catalysis Science Program and to the advancement of the field of catalysis. It is important that the balance of the technical focus of the initiative be maintained, including studies of high-surface-area catalysts, surface science, nanoscience, electrocatalysis, and theory.

BOX 5-2
Nanostructured Oxides: An Example of Impacts of the Catalysis Science Initiative on Fundamental Science

A challenge in solid-state catalysis is to understand the structure, composition, and functions of the families of catalytic sites normally present in solids. Although molecular bonding with the sites is local or of short range, extended interactions with the substrate, coadsorbents, and the various catalytic sites occur at longer ranges. As a result, reactivity is typically more difficult to control than in organometallic compounds, in which the sites can be well defined and isolated from one another. A recent trend in solid-state catalysis is the pursuit of single-site topologies, which are defined as noninteracting and uniform, or chemically identical, functions. Such an ambitious goal remains in general unachievable, but nanoscience may provide the tools to reach it.

Catalysis scientists over the past few years have dramatically improved their ability to design and synthesize inorganic sites with controlled size, atomic connectivity, and hybridization with either organic or other inorganic superstructures. Such solids contain chemical functions and physical properties that can be tuned for energy conversion, petrochemical synthesis, and environmental reactions. Many program-funded groups have made contributions to this area of fundamental research, such as Hrbek,[1] Iglesia,[2] Guliants,[3] Peden,[4] Suib,[5] and Wachs.[6]

[1] Rodriguez, J. A., S. Ma, P. Liu, et al. 2007. Activity of CeO_x and TiO_x nanoparticles grown on Au(111) in the water-gas shift reaction. *Science* 318(5857):1757-1760.

[2] Liu, H., and E. Iglesia. 2005. Selective oxidation of methanol and ethanol on supported ruthenium oxide clusters at low temperatures. *J. Phys. Chem. B* 109(6):2155-2163.

[3] Guliants, V. V., M. A. Carreon, Y. S. Lin. 2004. Ordered mesoporous and macroporous inorganic films and membranes. *J. Membr. Sci.* 235(1-2):53-72.

[4] Herrera, J. E., J. H. Kwak, J. Z. Hu, Y. Wang, C.H.F. Peden, J. Macht, and E. Iglesia. 2006. Synthesis, Characterization, and Catalytic Function of Novel Highly Dispersed Tungsten Oxide Catalysts on Mesoporous Silica. *J. Catal.* 239:200-211.

[5] Yuan, J. K., W. N. Li, S. Gomez, S.L. Suib. 2005. Shape-controlled Synthesis of Manganese Oxide Octahedral Molecular Sieve Three-dimensional Nanostructures. *J. Am. Chem. Soc.* 127(41):14184-14185.

[6] Wachs, I. E., Y. Chen, J. M. Jehng, L.E. Briand, T. Tanaka. 2003. Molecular Structure and Reactivity of the Group V Metal Oxides. *Catal. Today* 78(1-4):13-24.

FIGURE 5-4 SEM images of manganese oxide octahedral molecular sieves, which are a class of microporous transition metallic oxides. Manganese oxides are used extensively in chemical processes for ion-exchange, separation, catalysis, and energy storage in secondary batteries.
SOURCE: Yuan, J. K., W. N. Li, S. Gomez, S.L. Suib. 2005. Shape-controlled synthesis of manganese oxide octahedral molecular sieve three-dimensional nanostructures. *J. Am. Chem. Soc.* 127(41):14184-14185.

Theory

Electronic-structure calculations based on density functional theory (DFT) developed extremely rapidly in the 1990s to a point where the complex systems of interest in catalysis could be treated with high enough accuracy to become semiquantitative and have predictive power. That has changed the field, providing unprecedented insight into the details of surface bond-making and bond-breaking processes. It is now possible to calculate activation energies of elementary surface reactions for various reactions and catalysts and to understand trends in reactivity from one catalyst to the next (see, for instance, the work of Mavrikakis,[8] Barteau,[9] and Neurock.[10]). The theoretical methods can be used to describe more and more complex systems. Most recently, reactions at the solid–liquid interface of importance in electrochemistry have been treated by Neurock.[11] (Figure 5-4).

---

[8]Zhang, J., M. Vukmirovic, Y. Xu, Y.; Mavrikakis, M.; Adzic, R. R. *Angew. Chem. Int. Ed.* 44:2132.

[9]Linic, S., J. Jankowiak, M. A. Barteau. 2004. Selectivity Driven Design of Bimetallic Ethylene Epoxidation Catalysts from First Principles *J. Catal.* 224:489.

[10]Pallassana, V., and M. Neurock. 2000. Electronic Factors Governing Ethylene Hydrogenation and Dehydrogenation Activity of Pseudomorphic PdML/Re(0001), PdML/Ru(0001), Pd(111), and PdML/Au(111) Surfaces. *J. Catal* 191:301-317.

[11]Janik M. J., C. D. Taylor, and M. Neurock, 2009. First-principles Analysis of the Initial Electroreduction Steps of Oxygen Over Pt(111). *J. Electrochem. Soc.* 156:B126.

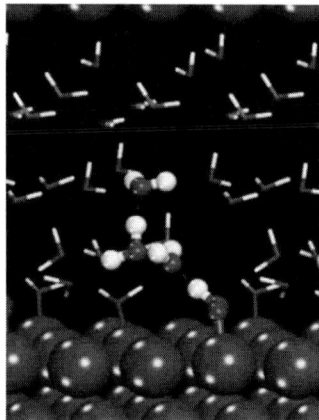

FIGURE 5-5 Transition state at 0.9 V for the first reduction of O_2 adsorbed to Pt(111) surface in water as determined by density functional theory. The label for each species is abbreviated by leaving the bulk solution phase 23 H_2O molecules off. Atoms shown in ball and stick format.
SOURCE: Janik M. J., C. D. Taylor, and M. Neurock. 2009. First-Principles Analysis of the Initial Electroreduction Steps of Oxygen Over Pt(111). *J. Electrochem. Soc.* 156:B126.

The advent of a quantitative theory of catalysis opens the door to the long-held dream of rational design of new catalysis. New design possibilities include the development of core-shell structures with new catalytic properties (Box 5-3) and metal alloys with improved selectivity (Barteau[12]).

Theory and computation have become an integral part of surface science and heterogeneous catalysis. They complement experiments, and many experimental groups include DFT calculations in their tools. Several industry representatives who were interviewed point to the development of theoretical methods as one of the major advances in heterogeneous catalysis over the past few years. Many industrial laboratories have separate theory groups.

Analysis of the Theory Subarea. BES responded immediately when new possibilities arose in the theory of heterogeneous catalysis and built up a strong portfolio: three grants were awarded from FY 1999 to FY 2001, six from FY 2002 to FY 2004, and 19 from FY 2005 to FY 2007. Six of the grantees received their funding early in their careers. The classification performed by DOE probably underestimates the growth of theory, especially in groups that are working with both experiment and theory.

---

[12]Linic, S., J. Jankowiak, and M. A. Barteau. 2004. Selectivity Driven Design of Bimetallic Ethylene Epoxidation Catalysts from First Principles. *J. Catal.* 224:489-493.

Theory grant holders are almost exclusively in universities; in the last funding period, only 1 of 19 principal investigators on the list worked in a national laboratory. The theory efforts have benefited from the supercomputing facilities that are available in the national laboratories, capital investment in the laboratories has had a rapid and important impact.

The Catalysis Science Program portfolio in heterogeneous catalysis theory is of a high international standard. The grant holders include most of the leading U.S. researchers in the field. Because the field is new, several principal investigators have received funding to develop new programs.

Contributions to Fundamental Science. The theory field is so new that it is clearly "cutting edge." Funded research is for both fundamental and applied science. It is fundamental in that it deals with a description of catalytic chemistry at the most basic level, and it is applied in that most of the projects are undertaken with a specific application (reaction or catalyst system) in mind.

Contributions to Energy Goals. Theory and computation will have a major influence on the development of heterogeneous catalysis in the future. They supplement experiments and offer new insight. They also offer new ways of building design tools that could point to new catalysts. The development of this field is essential for the development of new catalysts for energy transformations and more generally more efficient catalysts for the chemical industry.

The portfolio contains few projects related to the development of theoretical and computational methods. Essentially all of the support for theory is focused on heterogeneous catalysis; computational homogeneous catalysis and biocatalysis do not seem to enjoy the same support.

BOX 5-3
Near-Surface Alloys and Core-Shell Nanoparticles
An Example of Research with an Important Impact on Fundamental Science and Contributions to Meeting Energy Goals

The main objectives of closely combining theoretical modeling with experimental synthesis of nanomaterials are to arrive at first-principles design criteria that are consistent with the synthetic limitations and to advance synthesis and characterization methods to achieve the most revolutionary theoretical constructs. The work of Mavrikakis uses fundamental research and furthers fundamental understanding of mixed-metal alloys and near-surface alloys (NSAs) while developing catalysts for energy-relevant applications. It is a good example of fundamental work performed with the end in mind. Catalysis development for preferential oxidation of CO in hydrogen has focused on the use of these unique alloys with reliance on both experimental and theoretical techniques.[1] The work builds on previously funded BES catalysis research.[2]

SOURCE:
[1] Alagyouglu, S., A. U. Nilekar, M. Mavrikakis, and B. Eichorn. 2008. Ru–Pt core–shell nanoparticles for preferential oxidation of carbon monoxide in hydrogen. *Nature Materials* 7:333-338.

[2] Mavrikakis, M. and J. Greeley. 2004. Alloy Catalysts Designed from First-Principles. *Nature Materials* 3, 810-815.

Hydrogen Fuel Initiative

Analysis of the HFI Subarea. The first call for the HFI came out in the summer of 2004, and funding for it began in 2005. At that time, 10 proposals from researchers new to the Catalysis Science Program were funded (the HFI and the electrocatalysis program are lumped together here for convenience): 3 in national laboratories and 7 in universities. Many of the projects dealt with aspects of catalysis related to specific applications, such as catalysis for fuel cells or for reforming, but some were fundamental. For example, an HFI-funded project at Oak Ridge National Laboratory, titled "Nanoscale Building Blocks for Multi-Electron Electrocatalysis,"[13] aimed to develop a fundamental understanding of multielectron electrochemical reactions, an extremely important aspect of electrochemistry.

In 2007, three additional projects were funded, two in national laboratories and one in a university. One principal investigator was new to the program, one was already receiving funding, and one, who was not new to the program, had not been a principal investigator. The HFI funded a number of established catalysis researchers who had not been funded by BES (for example, U. Ozkan, A. Datye, and J. Rodriquez; see Appendix Table F-1). The new projects funded by BES under the HFI continued the BES effort to bring new researchers into the Catalysis Science Program.

Impact on Fundamental Science and Contributions to Meeting Energy Goals. Because the first year of funding was FY 2005 for most HFI and electrocatalysis proposals and FY 2007 for other projects, it is difficult to assess the impact of this body of work. However, the variety of electrocatalysis and catalysis research reflected in the portfolio is good, and much of the research results could be relevant to energy needs beyond the use of hydrogen as fuel. Some of the research reflects the technical challenges that arise when fuel hydrogen is produced from hydrocarbon resources (for example, CO poisoning on platinum electrodes and catalysts for reforming methane) rather than from electrolysis of water by solar or nuclear means. Future changes in energy policy could affect the relevance of this portfolio. Ideally, BES funding should be relatively immune to policy shifts.

HOMOGENEOUS CATALYSIS

Grants funded by the Catalysis Science Program in homogeneous catalysis and organometallic chemistry deal broadly with the design, synthesis, and

---

[13] Basic Energy Sciences. 2007. Frontiers in Interfacial and Nano Catalysis. U.S. Department of Energy and Oak Ridge Associated Universities. Online. Available at http://www.sc.doe.gov/BES/chm/Publications/Contractors%20Meetings/2007_Catalysis.pdf. Accessed January 13, 2008.

catalytic use of soluble compounds that have a set of ligands coordinated to one or more metals. The term *homogeneous catalysis* covers simple acid and base catalysis and enzyme catalysis, but this review considers only metal compounds. Homogeneous organometallic catalysts have long been used in industry, especially where high activity and selectivity are important. Homogeneous catalysis is a core field of chemistry in which the United States is the leader or among the leaders.[14] Recent Nobel prizes have been awarded for work involving olefin metathesis catalysis (Chauvin, Grubbs, and Schrock, 2005) and chiral hydrogenation and oxidation catalysis (Knowles, Noyori, and Sharpless, 2001). In addition, the survey of industry representatives (see Appendix D) identified a number of discoveries in homogeneous catalysis as *the most important breakthroughs in catalysis in the past 20 years.*

Why Is Homogeneous Catalysis Important for Energy?

Important reactions catalyzed by homogeneous catalysts include hydroformylation, olefin polymerization, C-H activation, cross-coupling catalysis, epoxidation, metathesis, asymmetric catalysis, and oxidation reactions. Such reactions, either in a homogeneous medium or through surface organometallic fragments in heterogeneous processes, give rise to many of the most important synthetic products of our time, including plastics, specialty and commodity chemicals, and pharmaceuticals.

Homogeneous catalysis can play a role in hydrogen production and storage and in the conversion of biobased feedstocks to fuels and chemicals. For example, soluble metal catalysts have recently been shown to convert bioderived glucose to 5-hydroxymethylfurfural (HMF) efficiently.[15] HMF has promise as a bioderived transportation fuel and as a source of monomers for production of advanced plastics. Homogeneous catalysts can also play a critical role in electrocatalysis, water-splitting catalysis, and artificial photosynthesis. The ability to design and control metal-containing catalysts that use abundant metals—such as iron, cobalt, and nickel—may be a critical part of any future hydrogen-based economy.

Homogeneous catalysts generally exhibit high selectivity, good yields, less waste, and excellent efficiency in the production of chemicals, polymers, and fuels. The high efficiency is especially important when the production of such materials involves petroleum or natural-gas-based feedstocks: greater efficiency results in the use of less fossil fuel as feedstock and of the fuel required in their production, separation, and transportation, and using less fossil fuel means reducing the generation of carbon dioxide. The high selectivity of homo-

---

[14]National Research Council. 2007. *The Future of U.S. Chemistry Research, Benchmarks and Challenges*. Washington, DC: National Academies Press.

[15]Zhao, H. J. E. Holliday, H. Brown, Z.C. Zhang. 2007. Metal Chlorides in Ionic Liquid Solvents Convert Sugars to 5-Hydroxymethylfurfural. *Science* 316:1597.

geneous catalysts can permit the conversion of simple molecules, such as methane, to fuels and chemical feedstocks and the conversion of crude oil, bioresources, and tar sands to useful fuels and small-molecule feedstocks. Nevertheless, because of the ease of separation of product from catalysts, engineers often prefer to use heterogeneous catalysts for technical applications.

A major strength of work in homogeneous catalysis is the ability to probe structure–function relationships and to translate the resulting information into the design of optimal catalysts. In general, the use of homogeneous catalysts makes it easier to understand new catalytic mechanisms. The fundamental understanding that results is often directly useful in the design of heterogeneous catalysts. As expressed by J. Basset at the most recent Catalysis Science Program contractor meeting,[16] the understanding of heterogeneous catalysts as proceeding via "surface organometallic fragments" may be used to design single-site catalysts on surfaces, with an ultimate goal of designing a new generation of hybrid catalysts whose improved reactions can be predicted from simple fundamentals of organometallic homogeneous chemistry.

Assessment of the Homogeneous Catalysis Portfolio

As is the case in the overall Catalysis Science Program portfolio, most of the grants for homogeneous catalysis were awarded to universities during the period FY 1999 to FY 2007 (Table 5-3). The 62 grants funded during the period FY 2005 to FY 2007 totaled approximately $30 million, which represented approximately 27 percent of total Catalysis Science Program funding.

The grants for work in homogeneous catalysis can be divided into several fields of research. Approximately one-half of the grants in FY 2007 involved C-H activation; the rest largely dealt with inorganic synthesis, single sites, and polymerization. Other topics covered were C-X activation, chiral–steroselective activation, aromatic reactions, and enzyme chemistry. Almost one-half of the principal investigators funded in FY 2005 to FY 2007 (22 of 50) had been funded in FY 1999 to FY 2001; approximately one-half were newly funded in FY 2005 to FY 2007, revealing a mix of support for established and for new investigators.

In this assessment of funding of work in homogeneous catalysis, single-site polymerization, C-H activation, homogeneous catalysis in organic synthesis, and homogeneous catalysis in biorelated projects are highlighted. These subjects are chosen particularly to demonstrate the development of the work of various investigators who participated in program contractor meetings.

---

[16] Basic Energy Sciences. 2008. Frontiers in Molecular Catalysis Science. U.S. Department of Energy and Oak Ridge Associated Universities. Online. Available at *http://www.sc.doe.gov/bes/chm/Publications/Contractors%20Meetings/2008_Catalysis.pdf*. Accessed January 13, 2008.

TABLE 5-3 Grants for Research in Homogeneous Catalysis (Including Biorelated Catalysts in Parentheses) in National Laboratories and Universities

	1999–2001	2002–2004	2005–2007
National laboratories	15 (1)	7	10
Universities	35	32 (4)	52 (7)
TOTAL	50 (1)	39 (4)	62 (7)

Single-Site Polymerization Catalysis

One of the most dramatic advances in catalysis during the past 25 years has been the development and understanding of well-defined single-site polymerization catalysts for the controlled production of commercially important polyolefin materials. The Catalysis Science Program has strongly supported single-site polymerization research from the inception of the field and must be credited with having a great impact on the development of the field. It is an outstanding example of the value of basic research in homogeneous catalysis that has advanced fundamental scientific understanding and has guided how we use energy and petrochemical resources more efficiently to affect everyday life.

Contributions to Fundamental Science. Early funding of fundamental research by the Catalysis Science Program provided the key understandings that were exploited to develop commercial products that use less energy, produce less waste, and are more recyclable than the products they replaced . For example, polyolefins, the primary products of single-site polymerization, are used to make products for the automotive, personal-care, clothing, durable-goods, and food industries.

The value of fundamental ligand design and mechanistic studies in homogeneous catalysis can be seen in the effect of research in homogeneous single-site olefin polymerization catalysis supported by BES. A new class of ligands was developed by principal investigator J. Bercaw and colleagues (Figure 5-6). The work involved a new class of mono-cyclopentadienyl silylamido complexes of scandium. The class was elaborated and combined with group 4 metals to produce highly active single-site polymerization catalysts that are now used in U.S. industry to produce over 2 billion pounds of polyolefins a year.[17] The new polyolefins include long-chain branched copolymers of ethylene with α-olefins, new elastomers, and a new process for EPDM rubber production.[18]

---

[17]McKnight, A. L., and R. M. Waymouth. 1998. Group 4 ansa-Cyclopentadienyl-Amido Catalysts for Olefin Polymerization. *Chem. Rev.* 98:2587-2598.

[18]Stevens, J. C., F. J. Timmers. D. R. Wilson, D. R.; Schmidt, G. F.; Nickias, P. N.; Rosen, R. K.; Knight, G. W.; Lai, S. Y. Eur. Patent Appl. EP 416815-A2, 1991 (Dow Chem. Co.); Canich, J. M. Eur. Patent Appl. EP 420436-A1, 1991 (Exxon Chem. Co.); Lai, S. Y., J. R. Wilson, G. W.

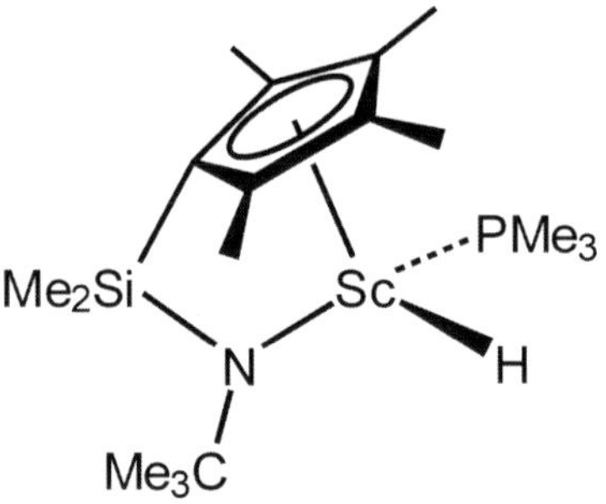

```
        ⬠
Me₂Si╲  │  ╲ ⋯PMe₃
       ╲ Sc╱
        N    H
        │
       Me₃C
```

FIGURE 5-6 New class of mono-cyclopentadienyl silylamido complexes of scandium, developed by J. Bercaw and colleagues.
SOURCE: Shapiro, P. J., E. Bunel, W. P. Schaefer, et al. 1990. Scandium Complex [{(.eta.5-C5Me4)Me2Si(.eta.1-NCMe3)}(PMe3)ScH]2: a Unique Example of a Single-component .alpha.-Olefin Polymerization Catalyst. *Organometallics* 9:867-869.

The power of breakthroughs in homogeneous catalysis can be seen in the new EPDM polymerization processes, which are so much more efficient, use so much less energy, and require so much less capital than prior technology that much of the world production of this important elastomer uses the new processes.[19] This important technologic achievement would have been hindered without the support of DOE funding.

It can be argued that new advances in science follow on the heels of new physical techniques. One way that the portfolio will contribute to meeting long-term national energy goals is through the discovery of cutting-edge analytic techniques for the study of catalytic mechanisms. C. Landis[20] has used funding from BES to develop new analytic techniques for monitoring homogeneous catalysis as substrates are turned into products and elucidating the mechanisms of organometallic reactions. To quote from Landis' abstract, "two general problems in catalysis that are particularly relevant . . . are (1) how much of the catalyst is active and (2) what are the rate laws for very fast processes?" The project shows how the portfolio will contribute to the advancement of fundamental sci-

---

Knight, G. W.; Stevens, J. C.; Chum, P.-W. S. (Dow Chem. Co.) U.S. Patent 5,272,236, 1993 (Dow Chem. Co.).

[19]Chum, P. S., W. J. Kruper, M. J. Guest. 2000. Materials Properties Derived from INSITE Metallocene Catalysts. *Adv. Mater.* 12:1759-1767.

[20] Basice Energy Sciences. 2006. Frontiers in Organometallic, Inorganic, and Bioinspired Chemistry & Catalysis. U.S. Department of Energy and Oak Ridge Associated Universities. Online. Available at
http://www.sc.doe.gov/bes/chm/Publications/Contractors%20Meetings/2006_Catalysis.pdf. Accessed January 12, 2009.

ence in energy-related catalysis. We mention it also to illustrate the cross-fertilization that is possible among the researchers who are funded by the program and who meet each other in the contractor meetings.

Landis and colleagues are developing two methods, stopped-flow nuclear magnetic resonance and quenched-flow mass spectrometry, that may revolutionize the ability of future scientists to study homogeneous catalysts as they are converting reactants to products. Critical features of the methods include the potential to distinguish active from inactive catalysts on millisecond time scales and the potential for automated, high-throughput operation."In principle, each method is capable of determining complete kinetic profiles very efficiently: in principle, it may be feasible to collect all the data needed to determine the rate laws of initiation, propagation, and termination for an industrially relevant catalyst in 15 min."

Contributions to Meeting Energy Goals. Research in single-site catalysis has contributed to our national goal of using less energy to produce the materials we need for our lives. And while the Catalysis Science Program has been supporting the discovery of important new classes of catalysts, it also has been supporting (see abstracts in the 2006 contractor meeting book[21]) detailed mechanistic investigations of centrally important initiation, propagation, and termination processes (J. Bercaw, R. Jordan, T. Marks, and R. Schrock); the nature of catalyst–cocatalyst interactions (T. Marks); the nature of surface-supported species (T. Marks); and techniques for determining the rate laws for very fast polymerization processes (C. Landis). With a view to gaining ever more control over polymer properties, tandem catalysts have been studied and developed (G. Bazan), and more recently studies of single-site polymerization catalysis have been extended to other industrially important polymers, such as polylactide and polycarbonates (G. Coates and M. Chisholm), including the development of a carbonylation route to lactone monomers (G. Coates).

Fundamental science that will continue to help us to reach our long-term energy goals has been developed. Ancillary benefits include avoiding the use of plasticizers, which are of increasing concern for potential environmental and long-term health effects. The program has undoubtedly made and will continue to make major contributions to the development of single-site polymerization catalysis.

---

[21] Basic Energy Sciences. 2006. Frontiers in Organometallic, Inorganic, and Bioinspired Chemistry & Catalysis. U.S. Department of Energy and Oak Ridge Associated Universities. Online. Available at http://www.sc.doe.gov/bes/chm/Publications/Contractors%20Meetings/2006_Catalysis.pdf. Accessed January 12, 2009.

C-H Activation

The Catalysis Science Program has a long history of funding projects in C-H bond activation and may rightly claim that it has made major contributions to successes in fundamental research.[22] The largest number of individual projects funded by the program are in C-H and C-X bond activation (see Table 3-2). The original "Holy Grail" in C-H activation, as expressed in the 1980s when this research began to flourish, was the conversion of methane gas into methanol, which was desirable as a liquid fuel for ease of transportation.

The productive functionalization of unactivated C-H bonds was first reported in Russia by A. E. Shilov in the 1960s as a noncatalytic process that was stoichiometric in expensive platinum reagents. At that time, observations of the elementary step of metal insertion into an aliphatic C-H bond, presumably the most difficult step, were few. In the ensuing decades, there were numerous demonstrations of C-H addition processes using redox active metals—primarily Ir(I), Ru(II), and Pt(II)—and mechanistic studies in selectivity of these metals toward hydrocarbon substrates.

Analysis of the C-H Activation Subarea. Despite much effort to date, for the portfolio to be effective in meeting national energy goals with respect to C-H activation, there must be further functionalization of the derived metal alkyl or metal aryl via known organometallic pathways in a catalytic manner. Nevertheless, it must be emphasized that 20 years ago chemists could not design catalysts to do what they now do routinely, that is, "crack" the unactivated C-H bond of simple hydrocarbons or even understand many of the principles that control selectivity of the process. Consistent with progress in the first stage of tapping into the reactivity of the "nonfunctional" C-H group is the large investment that has been made by the Catalysis Science Program portfolio, in which 34 of 140 projects (24 percent) are identified by the program directors in the 2007 projects as involving C-H activation (see Table 3-2).

The need for catalytic modification of the C-H group for fuel or for chemical-feedstock development will continue to be of interest and to challenge scientists and engineers. Until now, the program has focused its support on studies of the critical first step in C-H activation; however, there is a need to focus future research on the completion of important catalytic cycles to have an influence on energy problems of the future. The field continues to have great promise, but new approaches and new insights are needed to advance beyond single-metal activation of the C-H bond. The background for such studies should include related biological processes. As stated by Tobin Marks during his presentation to the committee (see Appendix C), we should "learn from Nature, then

---

[22]For a review of the history of C-H bond activation discoveries see: Goldman, A.S., and K. I. Goldberg. Organometallic C–H Bond Activation: An Introduction. In Activation and Functionalization of C-H Bonds ; Goldberg, K. I.; Goldman, A. S., Eds.; ACS Symposium Series 885; American Chemical Society: Washington, DC, 2004, pp 1-43.

go beyond." Although many projects are described as "bioinspired," few projects in the portfolio carefully analyze the mechanistic implications of enzyme active sites and the requirements met by the surrounding protein matrix (see the section on Biorelated Projects below). In contrast, there are examples of bioinspired approaches in the portfolio, including the subject of C-H activation. One of them is described below.

Contributions to Fundamental Science. A collaborative effort researching nanovessels between R. G. Bergman and K. Raymond, of the University of California, Berkeley, is funded by the Catalysis Science Program. Bergman's involvement began 25 years ago, when initial studies of a photochemically produced Cp*Ir(PMe$_3$) moiety led to reliable oxidative addition of C-H bonds from aliphatic and aromatic compounds. The extensive mechanistic work that he and his colleagues have carried out over the years has allowed the current project of Bergman and Raymond to develop (Figure 5-7).

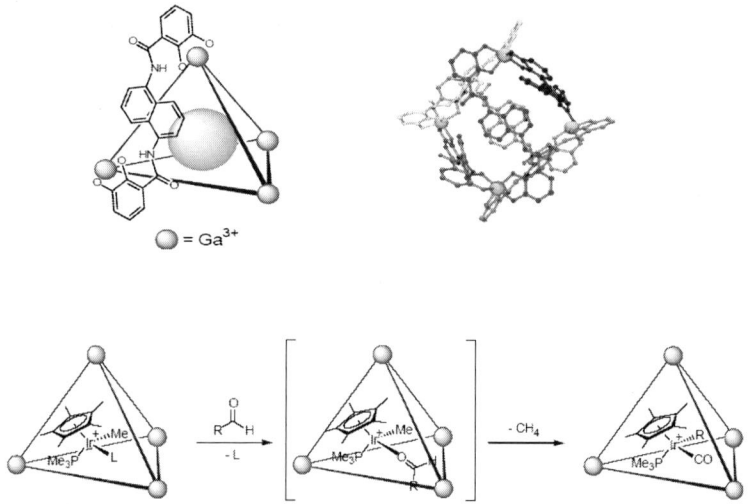

FIGURE 5-7 Top, "nanovessel" is composed of Raymond's supramolecular structure based on coordination of gallium by rigid binucleating dicatecholate ligand, which forms edges of tetrahedron. Bottom, cavity of water-soluble cluster is suitable for encapsulating Bergman's iridium complex, which further accepts substrate for reaction shown.
SOURCE: Raymond, K. N., and R. G. Bergman. Selective Organic and Organometallic Reactions in Water-Soluble Host-Guest Supramolecular Systems. In *Frontiers in Molecular Catalysis Science*, U.S. Department of Energy and Oak Ridge Associated Universities. Online. Available at http://www.sc.doe.gov/bes/chm/Publications/Contractors%20Meetings/2008_Catalysis.pdf. Accessed January 13, 2009.

Nanovessels are valuable because they provide high catalyst stability and selective substrate access. These nanoscale molecular "reactors" are composed of ligands that make the exterior hydrophilic, and leave the interior hydrophobic. This research can have major implications for the design of future semi-immobilized and site-isolated catalysts in other realms of homogeneous catalysis.

Mechanistic studies are the foundation of hypotheses whose testing requires new synthetic approaches. The extensive program of established Catalysis Science Program investigator W. Jones of the University of Rochester, demonstrates the necessity of these efforts, and the buildup of information regarding the interplay of catalyst structure and function has led to new results regarding selective processes in C-X functionalized hydrocarbons. Other grantees have developed the processes further for functionalization of the hydrocarbon. As shown in the most recent contractor meeting abstracts, approximately 12 of the reports by grantees addressed functionalization. Figure 5-8 shows some of the functionalization reactions.

Catalysis Science Program researcher M. Gagné has approached selective activation of C-X (X = O, Br, Cl) bonds in carbohydrates for conversion to new value-added products based on an abundant feedstock that is a renewable resource. His program also develops catalysts from a suitably ligated base metal (Ni) rather than from expensive noble metals.

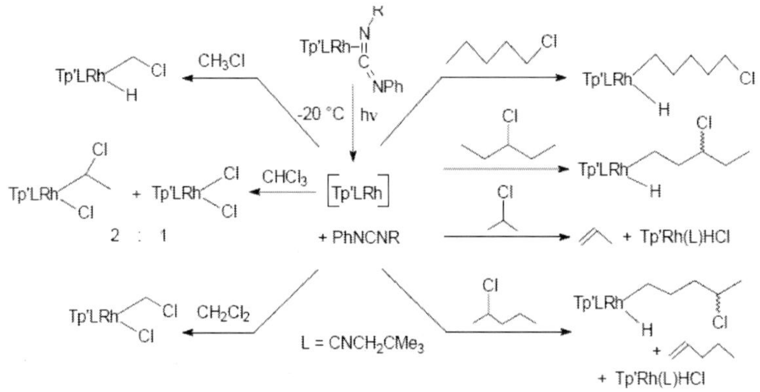

FIGURE 5-8 Activation of several chloroalkanes, in which it was observed that there is 100 percent preference for C-H activation of terminal methyl group over C-Cl bond.
SOURCE: Jones, W. 2008. Transition Metal Activation and Functionalization of Carbon-Hydrogen Bonds. In *Frontiers in Molecular Catalysis Science*. U.S. Department of Energy and Oak Ridge Associated Universities. Online. Available at
http://www.sc.doe.gov/bes/chm/Publications/Contractors%20Meetings/2008_Catalysis.pdf. Accessed January 13, 2009.

Principal investigator A. Goldman is showing leadership in several aspects of C-H activation, including dehydrogenation catalysis in tandem with alkylation (Figure 5-9). Such coupling of catalytic processes represents an important evolution from emphasis on C-H activation to hydrocarbon functionalization.

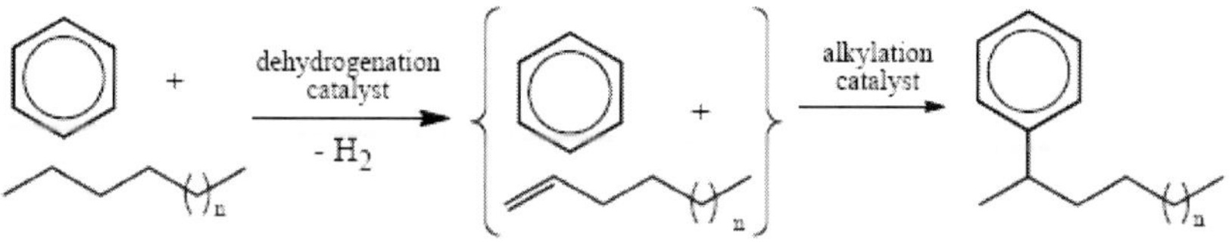

FIGURE 5-9 Tandem-catalyst system in which dehydrogenated products are subject to secondary reactions, such as addition of arenes (to yield alkylarenes) or cyclizations (to yield aromatics from linear alkanes).
SOURCE: Goldman, A. 2008. Transition Metal Activation and Functionalization of Carbon-Hydrogen Bonds. In *Frontiers in Molecular Catalysis Science*. U.S. Department of Energy and Oak Ridge Associated Universities. Online. Available at
http://www.sc.doe.gov/bes/chm/Publications/Contractors%20Meetings/2008_Catalysis.pdf. Accessed January 13, 2009.

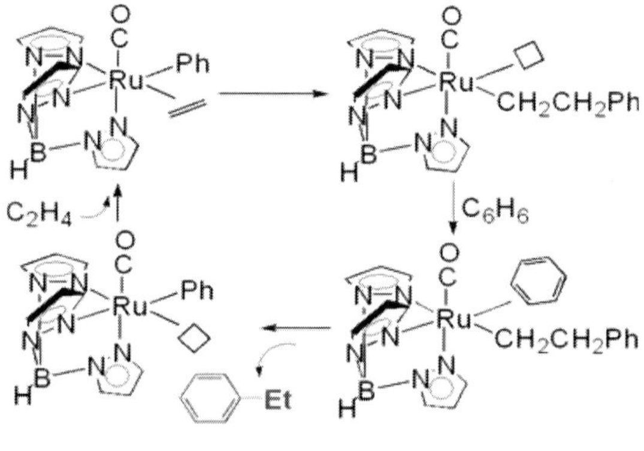

FIGURE 5-10 Proposed cycle of Ru(II)-catalyzed hydroarylation of olefins.
SOURCE: Gunnoe, T. B. 2008. Transition Metal Catalyzed Hydroarylation of Multiple Bonds: Exploration of Second Generation Ruthenium Catalysts and Extension to Copper Systems. In *Frontiers in Molecular Catalysis Science*, U.S. Department of Energy and Oak Ridge Associated Universities. Online. Available at
http://www.sc.doe.gov/bes/chm/Publications/Contractors%20Meetings/2008_Catalysis.pdf. Accessed January 13, 2009.

Another young investigator, B. Gunnoe, is addressing alkane functionalization by addition of C-H bonds across C-C multiple bonds (Figure 5-10). Both the C-H activation and the C-C coupling reactions appear to occur at a single ruthenium as a catalytic center.

Other catalysts, in particular planar complexes, have been shown to be promising for directing C-Cl versus C-H activation by yet another young investigator, O. Ozerov.

Contributions to Meeting Energy Goals. Ultimately, the goal of C-H activation catalysis is to find catalysts that would incorporate C-H activation into hydrocarbon-conversion technology, which would lead to functionalized compounds needed for feedstocks in the chemical industry or the ability to convert methane into useful liquid transportation fuels. Although the mechanistic studies of C-H activation processes that have established selectivity certainly encourage the development of possible applications, simple functionalization of hydrocarbons after C-H activation has not been realized. New ideas are needed; designs based on alkyl-group transfer to a second metal or on bifunctional ligands are possibilities.

The potential for C-H activation (and all other subfields of homogeneous catalysis) to affect future energy issues could be increased by integrating computational chemists more deeply into major synthetic and mechanistic studies. A fundamental issue in catalyst design and mechanistic understanding must be addressed by computational chemistry. There appears to be a deficiency of computational projects that address homogeneous catalysis in the Catalysis Science Program portfolio.

Homogeneous Catalysis in Organic Synthesis

Homogeneous catalysis is widely used in the synthesis of fine chemical, agricultural, and pharmaceutical intermediates and was identified by the survey of industry representatives (see Appendix D) as one of the most important fields of catalysis. It is a broad field, encompassing metal-based reagents for asymmetric and other transformations and metal-free organocatalyst systems. The discovery, development, and applications of metal-based catalytic reagents in organic synthesis have been recognized twice during the past decade: in 2001 by the award of a Nobel prize for chiral hydrogenation and oxidation catalysts and then in 2005 by the award of a Nobel prize for the olefin metathesis reaction.

Analysis of Homogeneous Catalysis in Organic Synthesis. Analysis of the portfolio shows only a few, but nevertheless important, grants in this area.

For example, the high inherent selectivity of homogeneous catalysts allows the production of molecules of desired handedness or enantioselectivity (asymmetric catalysis), which is critical for synthesis of fine chemicals, pharmaceuticals, agricultural chemicals, and electronic material.

An analysis of the portfolio reveals a small number of projects that involve the study and development of homogeneous palladium catalytic processes. Two projects in the portfolio deal with subjects of great interest. In the first, J. Hartwig uses palladium catalysis for hydroamination of olefins and addresses a key type of structural change. Wedding the strength of homogeneous catalysis to design a specific complex to affect a particular transformation with the strength of heterogeneous catalysts to facilitate separation of the metal complex from the reactants and products instills the need to "heterogenize" homogeneous catalysts. The other project, that of C. Jones, is a noteworthy effort to deal with that challenge.

Contributions to Fundamental Science. Asymmetric catalysis is of less importance for fuel production, but it is critical for production of fine chemicals, pharmaceuticals, agricultural chemicals, and electronics (for example, for liquid-crystalline displays). Conventional methods for separating enantiomers are slow and energy intensive. Asymmetric catalysts allow the production of selected enantiomers in high yields with increasingly efficient resource usage, energy efficiency, and waste reduction.

Given the historical and continuing importance of this field to the chemical industry and the fact that many of the organometallic and homogeneous catalysis programs funded by the Catalysis Science Program involve new ligand development, there seems to be a lack of programs directed specifically at asymmetric catalysis. Nevertheless, the number of these grants in the program portfolio is at an appropriate level considering that asymmetric catalysis research is for the most part well funded by other agencies.

Contributions to Meeting Energy Goals. The Catalysis Science Program funds very few grants dealing with the use of transition-metal catalysis with respect to synthetic organic chemistry, which has potential industrial and energy-related ramifications. For example, the catalytic properties of palladium could make it possible to functionalize simple hydrocarbon-based feedstocks to more complex molecules or to defunctionalize biobased feedstocks to target molecules for important applications.

Homogeneous Catalysis in Biorelated Projects

Enzymes are naturally occurring catalysts that are responsible for transforming biological molecules and materials into the myriad forms found in nature. Moreover, biological catalysts in energy-yielding and energy-requiring processes are linked to bioenergetics in ways that hold promise for meeting the energy needs of human economies.

An example is the cathodic reduction of oxygen to water—the key reaction of fuel cells. Engineered catalysts for fuel cells use noble metals, such as platinum. From a thermodynamic perspective, these are not efficient (and they are not naturally abundant). This has limited the use of fuel cells as power sources for transportation. Living cells have a version of the fuel cell, mitochondria, in which the cathodic reaction is the same as in engineered fuel cells: the reduction of oxygen to water to provide energy. Nature's catalysts for hydrogen oxidation (and for proton reduction to hydrogen) are typically nickel and iron or iron alone. Nature's catalysts smoothly oxidize methanol, ethanol, carbohydrates, and fats to carbon dioxide, collecting all of the electrons for use in the mitochondrial fuel cell to reduce oxygen and yield energy.

In addition to operating at ambient temperature, nature's catalysts generally direct complex chemical reactions along coordinates that yield essentially a single product, making the isolation and purification of products efficient. The protein component is a large fraction of an enzyme's structure and is necessary for biological functions. However, the chemical processes are carried out by a much smaller number of atoms in the enzyme than the protein, and are made up of abundant elements that actually control the movements of electrons, atoms, and ions to rearrange the reactants to make products. Research in bioinspired catalysis is focused on isolating and understanding those catalytically active components of the enzymes.

Analysis of Homogeneous Catalysis in Biorelated Projects. This portion of the Catalysis Science Program portfolio is difficult to assess because it is only a small component of the overall investment in catalysis. Depending on how one categorizes the funded projects, biorelated projects increased from 2 proposals during FY 1987 to FY 1997 to 10 during FY 2005 to FY 2007. The latest contractor meeting (May 2008) listed 14 projects that were described as bioinspired or biorelated.

Despite the importance of the biorelated projects, the program does not appear to provide a useful way to bring experts in biological catalysis into contact with the mainstream contributors to the catalysis portfolio. One of the more "biological" of the funded investigators is D. Kern. Her work in the dynamic motion of proteins is related to catalytic turnover and signal transduction (phosphorylation). A sampling of investigators whose primary aim is to understand metalloenzyme catalytic mechanisms is a needed component for advances in biorelated catalysis, and the lack of this opportunity for cross-fertilization is a serious concern. In contrast, many of the principal investigators who are active in bioinorganic catalysis are funded outside DOE. These principal investigators, however, may bring their insights to their DOE-funded research.

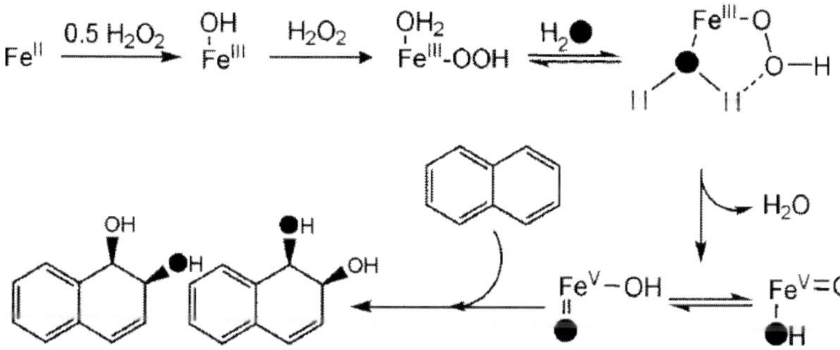

FIGURE 5-11 Proposed mechanism of bioinspired *cis*-dihydroxylation of naphthalene.
SOURCE: Que, L. 2008. Bio-inspired Iron Catalysts for Hydrocarbon Oxidations:
April 2008 report. In *Frontiers in Molecular Catalysis Science*, U.S. Department of Energy and Oak Ridge Associated Universities. Online. Available at *http://www.sc.doe.gov/bes/chm/Publications/Contractors%20Meetings/2008_Ca talysis.pdf.* Accessed January 13, 2009.

Contributions to Fundamental Science. The Catalysis Science Program has funded key researchers in biorelated catalysis with some success. We applaud the consistent emphasis on oxidation catalysis. For example, in a project titled "Bioinspired Iron Catalysts for Hydrocarbon Oxidations," L. Que and colleagues have improved the understanding of oxygen addition reactions to organic substrates (Figure 5-11). Inspired by nonheme iron oxygenases, Que and his group recently found biomimetic *cis*-dihydroxylation of naphthalene by using a six-coordinate, octahedral iron complex designed to have *cis*-oriented labile ligands. The proposed mechanism of reaction shown below is important for all *cis*-dihydroxylation reactions, and the specific application is important for bioremediation.

Because the selective oxidation of organic molecules is necessary for the efficient use of hydrocarbon feedstocks, it is important that this type of work be expanded. A second notable project in oxidation catalysis in the biorelated portfolio is that of the young investigator S. Stahl (see Appendix Table F-2), whose work on organometallic copper oxidase reactions is aimed at gaining fundamental understanding of copper-catalyzed aerobic oxidation reactions that proceed via organometallic intermediates. This new subject is being addressed by an up-and-coming young investigator who is moving away from expensive palladium reagents and toward the use of copper as a low-cost metal catalyst.

Another young investigator funded by the program is P. Chirik of Cornell University, whose work is in synthetic approaches to nitrogen fixation. The need for ammonia for fertilizer and for use in feedstocks in the production of nitrogen-containing organic molecules is as profound as are the energy requirements for abiological fixation. Chirik's metallocene complexes, which bind dinitrogen, promise new approaches to activation of this typically inert molecule.

It is notable that the detailed mechanistic approaches of L. Que and S. Stahl may be expected to also contribute to the understanding of the reversal of the oxygenation process. Oxygen removal from carbohydrates is important for the development of biofuels. The work of J. Dumesic of the University of Wisconsin includes two main approaches to reforming of oxygenated compounds (Box 5.1): biomass gasification, followed by water-gas shift and Fischer–Tropsch reaction, and hydrogenation of biomass to produce a liquid fuel.

The Catalysis Science Program staff has stated that a new emphasis on biomimetic chemistry will be announced in the near future. This will be highly appropriate given the convergence of molecular biology, biochemistry, biophysical techniques, protein crystallography and synthetic analogues of metalloenzyme active sites. Synthetic biology presents an opportunity for understanding the function of such evolutionarily perfected selective catalysts.

SUMMARY

On the basis of the information evaluated, the BES has done well with its investment in the Catalysis Science Program. Its investment has led to a greater understanding of the fundamental catalytic processes that underlie energy applications, and it has contributed to meeting long-term national energy goals by focusing research on catalytic processes that reduce energy use or explore alternative energy sources. In some areas the impact of the research has been dramatic, while in other areas important advances are yet to be made. The committee's key findings and recommendations for the Catalysis Science Program are summarized in Chapter 6.

6

Findings and Recommendations

After careful review of the Catalysis Science Program's research portfolio (grant titles, abstracts, individual researchers), especially for the fiscal years (FYs) 1999 to 2007, the committee concludes that the program has done well with its investment in catalysis basic research. The program's success can be attributed to key management decisions during the past eight years that have led to a current funding distribution that advances basic catalysis science in general and keeps the development of energy-related technologies developments as a long-term goal. The program has maintained support for many well-established and world-renowned leaders in catalysis, and, at the same time, has brought in many new researchers. The Catalysis Science Initiative (CSI) has been a particularly effective mechanism for bringing to the program new funds, new researchers, and innovative research topics—especially in heterogeneous catalysis. However, there are variations in the quality and relevance of the research in the program portfolio, as summarized in the committee's main findings and recommendations given below.

FINDINGS

The Catalysis Science Program portfolio is distributed between the two main categories of catalysis: heterogeneous and homogenous (see Table 3.1), each of which is assessed separately below. The committee has made this distinction for convenience, based on the traditional division in catalysis. However, researchers are increasingly crossing the traditional barriers between heterogeneous and heterogeneous catalysis, blurring the distinction between the two (see the discussion on Contractor Meetings in Chapter 4), which the committee views as a definite positive development.

Heterogeneous Catalysis

Heterogeneous catalysis accounts for the largest portion of the portfolio, and for the past eight years (FY 1999 to FY 2007), the program has made substantial progress in its support of the experimental and theoretical fundamental understanding of multiphasic (heterogeneous) catalytic systems, surfaces, and nanoscale structures. Contributions of the portfolio to national energy goals are also discussed where appropriate.

Traditional Heterogeneous Catalysis grants are awarded to individual investigators. These grants have been indispensable in establishing a long-term funding basis for several leading U.S. researchers in the field. The portfolio is highly important to research on energy efficiency and on alternative energy solutions. Pioneering work has been conducted in the areas of short-residence-time reactors; basic and acidic properties of catalysts using various probes and spectroscopic techniques; and aqueous-phase reforming of biomass for energy purposes.

Surface Science grants focus on achieving a better understanding of heterogeneous catalytic surfaces. Since its inception, the Catalysis Science Program has supported U.S. leaders in surface science and is now seeing a second generation of principal investigators, many of whom were graduate and postdoctoral students of the science's pioneers. During the past decade, the principal investigators in the surface science subarea have made numerous contributions to the mechanistic and structural understanding of catalytic reactions, which continue to advance catalysis of energy processes. Examples of this work include hydrogenation and dehydrogenation, reforming, selective oxidation, heteroatom removal, surface photochemistry and catalysis, structure and dynamics of catalyst surfaces, and bimetallic and alloy systems. The work is the foundation of the grand challenge to "Understand Mechanisms and Dynamics of Catalyzed Transformations," which is articulated in the recent report *Basic Research Needs in Catalysis for Energy* workshop.

Research and researchers funded by surface science grants also have contributed substantially to the growth of nanoscience and theory. Historically, much of heterogeneous catalysis and the research supporting it have been at the nanoscale. However, the increased and broader focus on nanoscience at the national level has changed the emphasis in surface science. During the most recent three-year period, approximately one-half of the projects focused primarily on surface reaction mechanisms, and the other half focused more on surface structure.

Nanoscience grants focus on emergent properties at the nanoscale. Funding for these grants began in 2001 as a result of the National Nanotechnology Initiative (NNI). Most of the NNI-funded work concentrates on the synthesis of novel single-site heterogeneous catalysts, nanoparticle catalysts, or new materials that might lead to a new family of catalysts. New materials are explored through new synthesis schemes that are used to make catalytic porous solids or by incorporating catalytic species into solid supports. Ten awards were originally funded under the NNI, and seven of them were still being funded in 2007. Overall, the new influx of funding for the Catalysis Science Program under the NNI has led to funding of several new investigators.

Catalysis Science Initiative (CSI) grants were first awarded in 2003 and were given to multi-investigator, multidisciplinary teams mainly involved in heterogeneous catalysis research. Few grants have been awarded for research in homogeneous catalysis and biocatalysis, despite the initiative's broader goal to develop "combined experimental and theoretical approaches to enable molecular-level understanding of catalytic reaction mechanisms." Although the 11 programs currently funded by the CSI are less than six years old, they already represent approximately 20 percent of the heterogeneous catalysis portfolio and have been successful in attracting and supporting investigators new to the field. This record suggests that the CSI has added value to the Catalysis Science Program and to the field of catalysis.

Theory grants are focused on theory, modeling, and simulation. Grants in other categories include theory but not as a main focus. Because the field is new, several grants have been used to build programs. The catalysis theory portfolio is considered to be of a high international standard. The list of grantees includes most of the leading U.S. researchers in the field. However, the current portfolio is somewhat lacking in the development of theoretical and computational methods, as well as in work focused on homogeneous catalysis and biocatalysis.

Hydrogen Fuel Initiative (HFI) grants focus on hydrogen production, storage, and use and involve mainly electrocatalysis. Many of the HFI-funded projects study the fundamental aspects of catalysis related to specific applications, such as catalysis for fuel cells or for reforming. Because the first year of funding was FY 2005 for most HFI electrocatalysis proposals and FY 2007 for other projects, it is difficult to assess the impact of this body of work. However, the collection of electrocatalysis and catalysis research in the portfolio is good. The research mostly reflects the technical challenges that arise when fuel hydrogen is produced from hydrocarbon resources (for example, carbon monoxide poisoning on platinum electrodes and the use of catalysts for reforming methane) rather than from electrolysis of water by solar or nuclear means. In addition, and similar to the CSI, the new HFI-funded projects have attracted new researchers to the Catalysis Science Program.

Homogeneous Catalysis

Grants for research in homogeneous catalysis constitute a smaller portion of the current Catalysis Program portfolio but have had an important impact on the Catalysis Science Program. For FY 2007, the grants were divided into two main research topics: approximately one-half involved C-H activation, and the other half involved mostly inorganic synthesis and inorganic single sites and polymerization. The committee also assessed the research topics of homogeneous catalysis in organic synthesis and in biorelated projects.

Single-Site Polymerization grants have made significant contributions to the understanding of fundamental catalysis. Single-site polymerization is one of the most important advances in catalysis of the past 25 years. The Catalysis Science Program has strongly supported this activity since its inception and must be credited with having a great impact on its development. This is an excellent example of the value of basic research and of how funding of productive, well-qualified individual principal investigators can lead to a successful commercial result of huge importance to chemical production and energy utilization.

C-H Activation and Functionalization grants have been a part of the Catalysis Science Program for a long time. The program has made major contributions to successes in fundamental research in this area. The ultimate goal of research in C-H activation catalysis is to find catalysts that will incorporate C-H activation into hydrocarbon-conversion technology, which will lead to functionalized compounds needed for feedstocks in the chemical industry or to the conversion of methane into useful liquid transportation fuels. However, the program has limited its impact by focusing its support on studies of only the first step in C-H activation. Simple functionalization of hydrocarbons after C-H activation has not been realized, and new ideas are needed. Designs based on alkyl group transfer to a second metal or on bifunctional ligands are possibilities. The study of C-H functionalization in biological processes also could help to inform research in this area.

Homogeneous Catalysis in Organic Synthesis grants are a very small but still important part of the Catalysis Science Program portfolio. For example, the high inherent selectivity of homogeneous catalysts allows the production of molecules of one desired handedness, or enantioselectivity (asymmetric catalysis), which is critical for the synthesis of fine chemicals, pharmaceuticals, agricultural chemicals, and electronic material. The selectivity of these catalysts presents the potential to conserve resources, increase energy efficiency, and reduce waste.

Biorelated grants are another small but important part of the Catalysis Science Program portfolio. Biological processes provide understanding of important

FINDINGS AND RECOMMENDATIONS

catalytic reactions such as C-H functionalization. Many projects in the homogeneous catalysis portfolio are described as bioinspired, but there are only a few examples of research that carefully analyzes the mechanistic implications of enzyme active sites and the requirements met by the surrounding protein matrix. Several of the program's principal investigators are active in bioinorganic chemistry but receive support for the work from other government agencies.

RECOMMENDATIONS

The Catalysis Science Program should continue its current approach to funding decisions. Multi-investigator and interdisciplinary programs such as the Catalysis Science Initiative should remain a part of the portfolio, but future teams might benefit from the inclusion of more homogeneous and biocatalysis researchers that are interested in energy solutions. The program should utilize future funding initiatives as a mechanism to maintain the balance of the program and to explore new approaches to carrying out research.

Influences on the Portfolio

The Catalysis Science Program should continue to broaden participation in its contractor meetings and other activities. Non-DOE sponsored workshop organizers and grantees funded by other BES programs should be invited to attend the Catalysis Science Program's activities to provide a more diverse influence on the portfolio. This is particularly important in the development of research directions that will have a long-term impact on the program.

Principal Investigators

The Catalysis Science Program should continue on its current path of maintaining support for productive, long-term researchers and of recruiting new researchers. The program also must ensure that the best researchers are identified and supported—this is especially important for heterogeneous catalysis because program funding is essential to the success of a heterogeneous catalysis researcher (see Chapter 3). The balance of funding for individual investigators and small groups should also be maintained.

Heterogeneous Catalysis

The distribution of the Catalysis Science Program's heterogeneous catalysis portfolio should be changed slightly. Studies on high surface area catalysts, surface science, nanoscience, and electrocatalysis should be maintained, but there should be a stronger emphasis on studies that explore catalyst design and new synthesis methods, unique reactor systems, unique characterization techniques, and completely new chemical reactions. Support for the development of theoretical methods also should feature more prominently in the portfolio.

Homogeneous Catalysis

A balanced homogeneous catalysis portfolio should extend beyond individual mechanistic steps to include greater development of new catalytic systems and reactions. The portfolio can be improved by pursuing opportunities in C-H bond functionalization, new approaches to transition-metal catalysis, and electrochemical catalysis (small molecule homogeneous catalysts supported on electrodes). In addition, there should be a greater emphasis on reducing highly oxidized compounds such as bioderived materials into fuels and feedstocks, and on bioinspired catalytic processes.

Appendix A

Statement of Task

At the request of the U.S. Department of Energy, Office of Science, Office of Basic Energy Sciences (BES), the National Academies shall review the investments by BES in catalysis science research.[1] The review will:

- Examine the BES research portfolio in catalysis and identify whether and how this portfolio has advanced fundamental science in this area.
- Discuss how the BES research portfolio in catalysis contributes and is likely to contribute to immediate and long-term national energy goals, such as reducing the Nation's dependence on foreign sources of energy.

---

[1] For the purposes of the review, BES catalysis research is defined as the grants portfolio funded in the Catalysis Science (formerly Catalysis and Chemical Transformations) critical research area.

Appendix B

Guest Speaker and Committee Biographic Information

GUEST SPEAKERS

Alexis T. Bell (NAE) is a professor of chemical engineering at the University of California, Berkeley, which he joined in 1967. Dr. Bell also has served as assistant dean and dean of the College of Chemistry and as chairman of the Department of Chemical Engineering. He is a faculty senior scientist in the E. O. Lawrence Berkeley National Laboratory. Dr. Bell is known for his research in heterogeneous catalysis and is recognized as one of the leaders in applying in situ spectroscopic techniques in combination with isotopic tracer techniques to the study of catalyzed reactions. Of particular note have been his investigations of the mechanism of Fischer–Tropsch synthesis, the synthesis of methanol, the selective catalytic reduction of nitric oxide, the oxidative dehydrogenation of alkanes, and the direct conversion of methane to oxygenates. He is the editor of *Catalysis Reviews* and *Chemical Engineering Science* and serves on the editorial boards of many other journals. He also has served on numerous committees of the American Chemical Society (ACS), the American Institute of Chemical Engineers (AIChE), the Council for Chemical Research (CCR), and the National Research Council. The results of his research have been published in more than 510 articles in refereed journals. Dr. Bell has received many honors for his research contributions, including the American Association of Engineering Education Curtis W. McGraw Award for Research; the AIChE Professional Progress, R. H. Wilhelm, and William H. Walker Awards; the North American Catalysis Society (NACS) Paul H. Emmett Award in Fundamental Catalysis and Robert Burwell Lectureship; the ACS Award for Creative Research in Homogeneous or Heterogeneous Catalysis; and the NACS and European Federation of Catalysis Societies Michel Boudart Award for the Advancement of Catalysis. He is a member of the National Academy of Engineering, a fellow of the American Association for the Advancement of Science, and a fellow of the American Academy of Arts and Sciences and has received an honorary professorship from the Siberian Branch of the Russian Academy of Sciences. He also has presented 11 named lectures at various universities and research institutes. Dr. Bell received his Sc.D. in chemical engineering from the Massachusetts Institute of Technology in 1967.

Charles P. Casey (NAS) is Homer B. Adkins Emeritus Professor of Chemistry at the University of Wisconsin-Madison. His research lies at the interface between organometallic chemistry and homogeneous catalysis, and his group studies the mechanisms of homogeneously catalyzed reactions. Dr. Casey is author of more than 250 papers in organometallic chemistry. He has served as chairman of the Organometallic Subdivision and of the Inorganic Chemistry Division of the American Chemical Society (ACS) and as president of ACS (in 2004), and he is a member of the editorial advisory board of the *Journal of the American Chemical Society*. In 1993, he was elected to the National Academy of Sciences and to the American Academy of Arts and Sciences. Dr. Casey received the ACS A.C. Cope Scholar Award in 1988 and Award in Organometallic Chemistry in 1991. He received his B.S. from St. Louis University and his Ph.D. from the Massachusetts Institute of Technology.

Michael J. Clarke was the program director for inorganic, bioinorganic, and organometallic chemistry at the National Science Foundation and holds a permanent position as a professor of chemistry at Boston College. His research focus is on how unusual transition-metal ions interact with biologic systems. He has designed and discovered new bioactive metal-containing agents for anticancer and other types of therapy, developed the activation-by-reduction hypothesis for metal anticancer agents, and participated in developing the concept that ruthenium anticancer compounds preferentially enter cancer cells by binding to transferrin. He was among the first to explore how ruthenium complexes bind to DNA and developed some of the early fundamental chemistry of technetium relevant to its use in radioimaging agents. He continues to explore how metal ions affect DNA, RNA, coenzymes, and important sulfur-containing polypeptides, such as glutathione. Dr. Clarke is interested in how nitrosyl ruthenium compounds can affect the strengthening of neuronal synapses through the release of nitric oxide at the neuronal site.

Anthony Cugini serves as director of the Department of Energy National Energy Technology Laboratory (NETL) Office of Research and Development (ORD), which comprises the on-site research personnel and laboratories in Morgantown, West Virginia; Pittsburgh, Pennsylvania; and Albany, Oregon. Before being named director of ORD, Dr. Cugini served as focus area leader of the NETL Computational and Basic Sciences Focus Area. During his tenure as focus area leader, NETL strengthened its position in computational research ranging from computational chemistry through larger-scale process modeling. Before coming to NETL in 1987, Dr. Cugini worked at Procter and Gamble and Gulf Research. At NETL, he has served primarily in ORD. Dr. Cugini has had a variety of research interests over a wide cross-section of energy and environmental technologies, including the kinetics of exothermic reactions, catalyst development, advanced carbon synthesis, hydrogen production and separation, gas hydrates, carbon dioxide sequestration, and computational modeling. His publications have included the topics of robotics technology, decontamination of military aircraft surfaces with novel polymeric materials, carbon dioxide sequestration technologies, novel catalysts, the effect of catalyst physical properties on activity, hydrogen separation and modeling, advanced carbon production, and computational modeling. Dr. Cugini received his B.S., M.S., and Ph.D. in chemical engineering from the University of Pittsburgh.

Robert J. Davis is a professor of chemical engineering at the University of Virginia, where he has served as the chair of chemical engineering since 2002. Dr. Davis has extensively used in situ spectroscopic methods coupled with both steady-state and transient kinetic methods to elucidate how oxide supports and basic promoters alter the active sites for a variety of catalytic reactions, including selective oxidation of hydrocarbons, acid–base reactions, and ammonia synthesis. He has received numerous awards, including the Emmett Award of the North American Catalysis Society, the National Science Foundation Young Investigator Award, the DuPont Young Professor Award, and the Union Carbide Innovation Recognition Award. He is the author or coauthor of about 100 publications, one patent, and a textbook titled *Fundamentals of Chemical Reaction Engineering*. Dr. Davis has served as president of the Southeastern Catalysis Society, chair of the 2006 Gordon research conference on catalysis, chair of catalysis programming of the American Institute of Chemical Engineers (AIChE), director of the Catalysis and Reaction Engineering Division of AIChE, and a member of the editorial boards of *Applied Catalysis A* and *B* and *Journal of Molecular Catalysis A*. He obtained his Ph.D. in chemical engineering from Stanford University in 1989 and was a postdoctoral research fellow in the Chemistry Department of the University of Namur in Belgium.

Bruce C. Gates (NAE) is a professor of chemical engineering at the University of California, Davis. He started his career as a research engineer at Chevron in 1967. In 1968, he joined the faculty of the University of Delaware, where he became the H. Rodney Sharp Professor of Chemical Engineering and served as director of the Center for Catalytic Science and Technology. In 1992, he moved to the University of California, Davis, where he is Distinguished Professor in the Department of Chemical Engineering and Materials Science. Dr. Gates has received numerous awards from the American Institute of Chemical Engineers and the American Chemical Society for his work in catalysis by metal clusters, solid acids, and zeolites. He is editor of *Advances in Catalysis*. Dr. Gates earned his B.S. in chemical engineering from the University of California, Berkeley in 1961 and his Ph.D. from the University of Washington in Seattle in 1966 and was a postdoctoral fellow in physical chemistry at the University of Munich.

L. Louis Hegedus (NAE) retired in 2006 after 10 years of service as the senior vice president for research and development of Arkema Inc., a diversified chemical company headquartered in Paris. He was responsible for all R&D in North America and for R&D coordination between the United States and France. His previous positions include 16 years with W. R. Grace, where he was a research vice president for specialty chemicals, and 8 years with the General Motors Research Laboratories, where he contributed to the development of the catalytic converter for automobile emission control. Dr. Hegedus is a past chairman of the Chemical Engineering Section of the National Academy of Engineering and a past chairman of the Council for Chemical Research. He received his Ph.D. in chemical engineering from the University of California, Berkeley and an M.S. in chemical engineering from the Technical University of Budapest.

W. Christopher Hollinsed is director of the Office of Research Grants of the American Chemical Society (ACS), which includes the ACS Petroleum Research Fund. The Petroleum Research Fund, with current assets of over $600 million, has provided funding for advanced scientific education and fundamental research in the petroleum field for over 52 years. Dr. Hollinsed joined ACS in 2005 after 26 years in the corporate world in a variety of research and leadership positions at DuPont and Polaroid. At DuPont, he was manager of academic programs, heading DuPont's Young Professor grant program and the Science and Engineering grants program. He received his Ph.D. from the University of Wisconsin and his B.S. from City College of New York. He is a fellow of the American Association for the Advancement of Science and recently received a Service Award from the National Organization for the Professional Advancement of Black Chemists and Chemical Engineers for serving as an advocate for the organization.

Gretchen Jordan is a principal member of the technical staff of Sandia National Laboratories. Since 1993, she has worked with the U.S. Department of Energy (DOE) to develop innovative methods for assessing the effectiveness of research programs. Projects with the DOE Office of Science include methods to assess and improve the research environment and identification of best practices in the management of science. She also works with DOE Energy Efficiency and Renewable Energy offices on evaluation and performance measurement at the project, program, and portfolio levels and assists the Sandia Science and Technology Strategic Management Unit in those activities. She has edited two special issues on measuring R&D performance in the *Journal of Technology Transfer* (July 1997) and *Evaluation and Program Planning* (1999) and contributed chapters to two books on evaluating science program. Before joining Sandia, Dr. Jordan was chairman of the Business Administration Department at the College of Santa Fe and a member of the staff of Senator Pete V. Domenici and the Senate Budget Committee.

Tobin J. Marks (NAS) is the Charles E. and Emma H. Morrison Professor and Vladimir N. Iptieff Professor of Chemistry at Northwestern University. Through landmark synthetic, mechanistic, and thermodynamic investigations, he and his students opened a new portion of the periodic table to organometallic chemistry. He has also made major advances in solid-state, polymer, bioinorganic, and boron hydride chemistry and in photochemical isotope separation. During his career, Dr. Marks has received numerous awards. Recent honors include the 2005 National Medal of Science, the American Institute of Chemists Gold Medal, the John C. Bailar Medal from the University of Illinois at Urbana-Champaign, the Sir Edward Frankland Prize Lectureship of the British Royal Society of Chemistry, and the Karl Ziegler Prize of the German Chemical Society. Dr. Marks is also the recipient of three American Chemical Society (ACS) national awards and the ACS Chicago Section's 2001 Josiah Willard Gibbs Medal. He was elected to the National Academy of Sciences and the American Academy of Arts and Sciences in 1993 and to the German Academy of Natural Scientists Leopoldina in 2005. He received his B.S. from Maryland University and his Ph.D. from the Massachusetts Institute of Technology.

Raul Miranda is a program manager in the U.S. Department of Energy Catalysis and Chemical Transformations Program and has 17 years of academic experience at the University of Louisville as a professor of chemical engineering. In 1990–1991, he was a visiting professor at the University of Mar del Plata in Argentina and Ecole Nationale Sup. de Chimie in Montpellier, France. From 1984 to 1989, he was summer research faculty associate at the Argonne National Laboratory in the Chemistry Division and the Materials Science Division. From 1996 to 1999, Dr. Miranda was program director of the Kinetics and Catalysis Program at the National Science Foundation. His teaching encompassed traditional chemical engineering courses, emphasizing graduate-level kinetics and reaction engineering, heterogeneous catalysis, engineering mathematics, solid-state chemical processing, and computational condensed-matter chemistry. His research interests are in catalytic-reaction mechanism identification. Dr. Miranda studied hydrogenation catalysis over self-assembled chirally modified surfaces, partial oxidation of alcohols and aldehydes over transition-metal oxide nanoparticles, and hydrotreatment of N-heteroaromatics over acidic supported transition-metal oxides. He is also interested in chemical microdevice technology, in particular the fabrication and characterization of solid-state microsensors. Dr. Miranda received his B.S. from the University of Cuyo in Argentina and his M.S. and Ph.D. from the University of Connecticut in Storrs.

John Regalbuto is the director of the Catalysis and Biocatalysis Program in the Directorate for Engineering in the National Science Foundation. Dr. Regalbuto's education includes a B.S. in chemical engineering from Texas A&M University in 1981, and an M.S. in chemical engineering and a Ph.D. from the University of Notre Dame in 1983 and 1986, respectively. Directly thereafter, he joined the University of Illinois at Chicago, his home institution, where he is a professor in the Department of Chemical Engineering. Dr. Regalbuto has several hundred research publications and presentations and most recently edited one of the few books in his research specialty, catalyst preparation. He has twice served as president of the Catalysis Club of Chicago and has been active in organizing symposia on catalysis for meetings for the American Institute of Chemical Engineers and the American Chemical Society.

Douglas Ray is interim deputy director for science and technology and associate laboratory director in the Fundamental & Computational Sciences Directorate of the Pacific Northwest National Laboratory (PNNL). As deputy director, Dr. Ray is responsible for guiding the laboratory's overall capability-development strategies, defining and advancing its science and technology portfolio, coordinating its scientific discretionary investments, and integrating its science and technology base to deliver essential scientific capability and accomplishments that advance the Department of Energy's missions. As associate laboratory director, he is responsible for PNNL's Office of Science and National Institutes of Health research programs and directs more than 500 staff members in four research divisions: Atmospheric Sciences and Global Climate Change, Biological Sciences, Chemical & Materials Science, and Computational Sciences and Mathematics. Dr. Ray joined PNNL in 1990. He earned a B.S. in physics from Kalamazoo College and a Ph.D. in chemistry from the University of California, Berkeley.

COMMITTEE

Nancy B. Jackson (Co-chair) is the manager of the International Chemical Threat Reduction Department of Sandia National Laboratories (a National Nuclear Security Administration laboratory managed and operated by Lockheed Martin). Dr. Jackson founded the department as the first partner of the U.S. Department of State to develop the Chemical Security Engagement Program, a scientific engagement program designed to raise the awareness of chemical security and safety among chemical practitioners around the world. At Sandia, Dr. Jackson previously held the positions of deputy director of the International Security Center; manager of the Chemical and Biological Sensing, Imaging & Analysis Department; and member of the technical staff leading heterogeneous catalysis research. Her technical experience lies primarily in imaging and in structure–property relationships of catalytic materials. Dr. Jackson earned her bachelor's degree in chemistry from George Washington University and her Ph.D. in chemical engineering from the University of Texas, Austin. She is a fellow of the American Association for the Advancement of Science and recently served as a member of the Board of Directors of the American Chemical Society.

Jens K. Nørskov (Co-chair) is professor of physics at the Technical University of Denmark (DTU). In addition he serves as director of the Lundbeck Foundation's Center for Atomic-scale Materials Design, chairman of the Nanotechnology Center at DTU, and chairman of the Danish Center for Scientific Computing. Dr. Nørskov's research interests span the theoretical description of surfaces, catalysis, materials, nanostructures, and biomolecules. He received his M.Sc. in physics and chemistry in 1976 and his Ph.D. in theoretical physics in 1979 from Aarhus University, Denmark. After experience at the IBM T.J. Watson Research Center in Yorktown Heights, New York, and the Nordic Institute of Theoretical Physics, Copenhagen, he joined the staff of Haldor Topsøe, Lyngby, Denmark, in 1985. In 1987, he received a special appointment as research professor by the Danish minister of education and joined DTU, where he was appointed professor of physics in 1992. Dr. Nørskov has held visiting professorships at the University of California Santa Barbara and the University of Wisconsin, Madison. He has received a number of awards, and is a member of Royal Danish Academy of Science and letters and of the Danish Academy of the Technical Sciences.

Mark A. Barteau (NAE) is senior vice provost for research and strategic initiatives and Robert L. Pigford Chair of the Department of Chemical Engineering of the University of Delaware. He received his B.S. in chemical engineering from Washington University in St. Louis and his M.S. and Ph.D. from Stanford University. He was a National Science Foundation postdoctoral fellow at the Technische Universität München before joining the University of Delaware faculty as an assistant professor of chemical engineering and associate director of the Center for Catalytic Science and Technology in 1982. He became director of the Center for Catalytic Science and Technology in 1996. In 2000, Dr. Barteau became the chairperson of the Department of Chemical Engineering. He also has held visiting appointments in chemical engineering at the University of Pennsylvania and in chemistry at the University of Auckland, New Zealand. Dr. Barteau's research, presented in more than 200 publications and a similar number of invited lectures, focuses on chemical reactions at solid surfaces and their applications in heterogeneous catalysis. He was one of the pioneers in demonstrating the application of surface spectroscopy to study the mechanisms of organic relations on single-crystal metal oxide surfaces, and such studies remain an important component of his research today. Dr. Barteau is the recipient of numerous awards, including the inaugural International Catalysis Award, presented by the International Association of Catalysis Societies in 1998; the 1995 Ipatieff Prize from the American Chemical Society; the Paul H. Emmett Award in Fundamental Catalysis, given by the North American Catalysis Society; and the 1993 Canadian Catalysis Lecture Tour Award of the Catalysis Division of the Chemical Institute of Canada. He has served as associate editor of the *AIChE Journal* and on the editorial boards of a number of other journals, including the *Journal of Catalysis*. He was a member of the National Research Council committee that produced the report *Beyond the Molecular Frontier: Challenges for Chemistry and Chemical Engineering.* He was elected to the National Academy of Engineering in 2006.

Mark J. Cardillo is the executive director of the Camille and Henry Dreyfus Foundation. Dr. Cardillo received his B.S. from Stevens Institute of Technology in 1964 and his Ph.D. in chemistry from Cornell University in 1970. He was a research associate at Brown University, a CNR research scientist at the University of Genoa, and a PRF research fellow in the Department of Mechanical Engineering of the Massachusetts Institute of Technology. In 1975, Dr. Cardillo joined Bell Laboratories as a member of the technical staff in the Surface Physics Department. He was appointed head of the Chemical Physics Research Department in 1981 and later named head of the Photonics Materials Research Department. Most recently, he held the position of director of broadband access research. Dr. Cardillo is a fellow of the American Physical Society. He has been a Phillips Lecturer at Haverford College and a Langmuir Lecturer of the American Chemical Society. He received the Medard Welch Award of the American Vacuum Society in 1987, the Innovations in Real Materials Award in 1998, and the Pel Associates Award in Applied Polymer Chemistry in 2000.

Marcetta Y. Darensbourg is a professor of chemistry at Texas A&M University. She received a B.S. from Union College Kentucky in 1963 and a Ph.D. from the University of Illinois at Urbana in inorganic chemistry in 1967. Her research focus is synthetic and mechanistic inorganic chemistry, including functioning models of catalytic active sites in bioinorganic–organometallic systems (nickel, iron, and cobalt). She was formerly a professor of inorganic chemistry at Tulane University and chair of the Inorganic Division of the American Chemical Society (ACS) from 1988 to 1990. She has received the ACS Distinguished Service in Inorganic Chemistry Award, the Association of Former Students Teaching Award, and the Association of Former Students Research Award.

Anne M. Gaffney is vice president of the Technology Development Center of Lummus Technology. Dr. Gaffney is involved in activities related to developing catalysts for use in the oil and gas, petroleum-refining, and petrochemical-process industries. Gaffney received bachelor's degrees in chemistry and mathematics from Mount Holyoke College and a Ph.D. in physical organic chemistry from the University of Delaware. She has been involved in industrial chemistry and chemical engineering for 25 years, contributing to the technology portfolios of Arco Chemical Company, DuPont, and Rohm and Haas, in addition to Lummus Technology.

APPENDIX B

Vernon C. Gibson is the Sir Edward Frankland BP Professor of Inorganic Chemistry at Imperial College. He studied for his D.Phil. under Malcolm Green at the University of Oxford and then spent two years as a NATO postdoctoral fellow with John Bercaw at the California Institute of Technology. He heads the catalysis and materials research section in the Chemistry Department at Imperial, where his main research interests have been the design of new catalyst systems for the controlled synthesis of a number of industrially important classes of polymer. Dr. Gibson is a fellow of the Royal Society of London. He has recently taken up an appointment as Chief Chemist at BP, where he is responsible for providing strategic guidance and overview of the company's chemical activities.

Sossina M. Haile is a professor of materials science and chemical engineering at the California Institute of Technology (CIT). She received her B.S. and Ph.D. (1992) from the Massachusetts Institute of Technology and an M.S. from the University of California, Berkeley. Before joining the CIT faculty in 1996, Dr. Haile spent three years as an assistant professor at the University of Washington, Seattle. Dr. Haile's research centers on ionic conduction in solids with the twin objectives of understanding the mechanisms that govern ion transport and applying the understanding to the development of advanced solid electrolytes and novel solid-state electrochemical devices. Technologic applications of fast-ion conductors include batteries, sensors, ion pumps, and fuel cells. It is to the latter that Dr. Haile's work is most closely tied.

Masatake Haruta is a member of the faculty of urban environmental sciences of Tokyo Metropolitan University. He graduated from Nagoya Institute of Technology with a major in industrial chemistry in 1970. He received his Ph.D. in industrial chemistry from Kyoto University in 1976; his work there concerned electrochemistry in hydrogen fluoride solvent. In 1976, he was employed as a research scientist by Osaka National Research Institute (ONRI), where he was involved in the catalytic combustion of hydrogen. He studied catalyst preparation through colloid chemistry under Bernard Delmon's guidance at the Université Catholique de Louvain in Belgium in 1981–1982. Since his return to ONRI, he has been studying the catalysis of gold nanoparticles. In 1994, he was promoted to head of the Catalysis Section, a science fellow, and head of the Interdisciplinary Basic Research Laboratory. Dr. Haruta became director of the Department of Energy and the Environment in 1999. On the occasion of the transformation of national research institutes into semiautonomous bodies in 2001, he moved to Tsukuba as director of the National Institute of Advanced Industrial Science and Technology (AIST) Research Institute for Green Technology. In 2005, he joined the Faculty of Urban Environmental Sciences of Tokyo Metropolitan University. During his career at AIST, he was a guest professor at the Technical University of Vienna in 1994. He was also an adjunct professor of Osaka University from 1996 to 2005. Dr. Haruta pioneered a novel field of heterogeneous catalysis with the discovery of the unique catalytic performance of gold nanoparticles deposited on transition-metal oxides. He received the Osaka Science Prize, the Catalysis Society of Japan Science Award, and the International Precious Metals Institute Henry J. Albert Award.

Nenad M. Markovic is a principal investigator and senior scientist at the Argonne National Laboratory (ANL). He also is a group leader in the catalyst-development program for energy conversion and storage systems. Before joining ANL in 2005, he was a staff scientist at the Lawrence Berkeley National Laboratory from 1991 to 2005 and a group leader at the Institute of Electrochemistry of the University of Belgrade, Serbia, from 1986 to 1991. Dr. Markovic received his B.Sc., Ms.D., and Ph.D. in technology engineering from the University of Belgrade. He was one of the pioneers in surface electrochemistry on well-characterized single-crystal surfaces and in the use of surface science to develop electrocatalysts. He now combines electrochemical methods with ex situ ultra-high-vacuum spectroscopy and in situ surface x-ray scattering, scanning tunneling microscopy, and surface vibrational spectroscopy techniques in focusing on synthesis of anode and cathode catalysts for fuel-cell reactions, metal-deposition processes, and electrochemistry on transition-metal oxides. He is the author or coauthor of more than 200 papers and U.S. patents.

Thomas A. Moore is a professor of chemistry and biochemistry at Arizona State University (ASU) and director of the Center for the Study of Early Events in Photosynthesis in the College of Liberal Arts and Sciences. He is the interim director of the Center for Bioenergy and Photosynthesis in the Global Institute of Sustainability at ASU. He was awarded a Chaire Internationale de Recherche Blaise Pascal, Région d'Ile de France, Service de Bioénergétique, CEA Saclay, France, for the period 2005–2007. Dr. Moore has a Ph.D. in chemistry from Texas Tech University. He served as president of the American Society for Photobiology in 2004 and received the Senior Research Award from the society in 2001. He teaches undergraduate and graduate courses in biochemistry at ASU and lectures in biophysics at the Université de Paris Sud, Orsay. Dr. Moore's research in artificial photosynthesis is aimed at the design, synthesis, and assembly of bioinspired constructs capable of sustainable energy production and use.

Brendan D. Murray is a senior member of the Catalysts Department at Shell Oil Company. In addition, Dr, Murray is often asked to work closely with joint-venture partners and select global customers. In his 22 years at Shell, he has been involved in a number of commercial developments in the petrochemical and refining fields. His most important contributions have been in zeolite catalysis, novel catalytic processes, surfactants, and difficult separations. In 2007, Dr. Murray served as co-chair of the 20th North American Catalysis Society Meeting in Houston.

James C. Stevens is a research fellow in Core Research and Development at the Dow Chemical Company, where he has worked for 28 years. His primary field of research is new catalysts, particularly polyethylene, polypropylene, ethylene–styrene copolymers, and organometallic single-site catalysts. Dr. Stevens has been involved in the discovery and commercial implementation of Dow's INSITE technology and constrained-geometry catalysts, which are used in the production of over 2 billion pounds of polyolefins per year. He is an inventor on 82 issued U.S. patents, has 16 publications, and is the editor of one book. He has won a Dow Inventor of the Year Award five times and was presented the Dow Central Research Excellence in Science Award. In 1994, Dr. Stevens was a co-recipient of the U.S. National Inventor of the Year Award. In 2002, Dow was awarded the National Medal of Technology by President George W. Bush, in part on the basis of Dr. Stevens's work in olefin polymerization catalysis. He is the 2004 recipient of the American Chemical Society (ACS) Delaware Section Carothers Award, which honors scientific innovators who have made outstanding advances in and contributions to industrial chemistry. He was awarded the ACS Award in Industrial Chemistry in 2006. Dr. Stevens also received the Herbert H. Dow Medal, the highest honor that Dow bestows on the company's scientists and researchers. He recently received the 100th Perkin Medal, widely considered to be the highest honor in American industrial chemistry. He was the 2007 recipient of the University of Chicago Bloch Medal. Dr. Stevens received a B.A. in chemistry from the College of Wooster in 1975 and a Ph.D. in inorganic chemistry from Ohio State University in 1979.

Barry M. Trost (NAS) is the Tamaki Professor of Humanities and Sciences in the Stanford University Department of Chemistry. His research interests include organic synthesis, catalysis, insect hormones and pheromones, antibiotics, antitumor agents, organic conductors, and the chemistry of sulfur, selenium, silicon, tin, palladium, and molybdenum. He received his undergraduate degree from the University of Pennsylvania in 1962 and a Ph.D. in organic chemistry from the Massachusetts Institute of Technology.

Appendix C

Guest Speakers

MEETING 1, January 10, 2008

The Petroleum Research Fund Review Process and Current Status
    W. Christopher Hollinsed, Director, ACS Petroleum Research Fund

Evaluating Research Funding Programs
    Gretchen Jordan, Sandia National Laboratory

NAS Review of the BES Catalysis Science Program (Study Sponsor Presentation)
    Eric Rohlfing, Director, Chemical Sciences, Geosciences, and Biosciences Division

Catalysis Science Program: Chemical Transformations Team
    Raul Miranda, Program Manager, Catalysis Science

Basic & Applied Catalysis Research for Hydrogen Production, Storage, and Fuel Cells
    JoAnn Milliken, Hydrogen Program, Energy Efficiency and Renewable Energy (EERE)

National Energy Technology Laboratory—Office of Research and Development
    Anthony Cugini, Fossil Energy, Fossil Energy Program, National Energy Technology Laboratory

Catalysis Research for Efficient and Renewable Energy
    Brian Valentine, Industrial Technologies Program, EERE

National Science Foundation Chemistry Division Catalysis Basic Research Funding

Michael Clarke, Inorganic, Bioinorganic and Organometallic Chemistry, Division of Chemistry

Catalysis and Biocatalysis at NSF and Its Relationship to Catalysis at DOE/BES
John Regalbuto, Catalysis and Biocatalysis, Division of Chemical, Bioengineering, Environmental, Transport Systems

MEETING 2, March 17, 2008

Catalysis and the Future of U.S. Chemistry: Benchmarks and Challenges
Charles Casey, University of Wisconsin
Tobin Marks, Northwestern University

Catalysis and Benchmarking the Research Competitiveness of U.S. Chemical Engineering
Louis Hegedus, retired, Arkema

International Assessment of Research in Catalysis by Nanostructured Materials
Robert Davis, University of Virginia

BES Catalysis Program
Raul Miranda, U.S. Department of Energy

Directing Matter and Energy: Five Challenges for Science and the Imagination: Overview of the BESAC "Grand Challenges" Report: Relevance to Catalysis Research
Tobin Marks, Northwestern University

Basic Research Needs in Catalysis for Energy Workshop
Alexis Bell, University of California, Berkeley, Lawrence Berkeley National Laboratory
Bruce Gates, University of California, Riverside
Douglas Ray, Pacific Northwest National Laboratory

Appendix D

Industry Questionnaire Respondents

Name	Affiliation
Armor, John N.	Air Products and Chemicals, Inc. (retired)
Bare, Simon	UOP
Bryndza, Henry	E. I. du Pont de Nemours & Company
Clausen, Bjerne	Haldor Topsøe
Daage, Michel	ExxonMobil Corporation
Ellis, Paul	Saudi Basic Industries Corporation (SABIC)
Farrauto, Robert	BASF/Engelhard Corporation
Francz, Thaddeus	Abbott Laboratories
Fugita, Terunori	Mitsui Chemicals
Green, Mike	Sasol
Groten, Will	Catalytic Distillation Technologies (CDTech)
Harth, Klaus	BASF
Lennon, Ian	Dow ChiroTech Technology Ltd.
Maughon, Robert	The Dow Chemical Company
McCabe, Bob	General Motors
McNally, John	Ineos
Monnier, John R.	Eastman Chemical Company
Moore, Eric	Ineos
Ovalles, Cesar	Chevron Corporation
Razavi, Abbas	Total
Schinski, William	Chevron Corporation
Sell, Thorsten	Novolen
Senanayake, Chris	Boehringer-Ingelheim
Soled, Stuart	ExxonMobil Corporation
Stiltz, Ulrich	Sanofi-Aventis
Sun, Yongkui	Merck & Co., Inc.
Sunley, Glenn	BP
Topsøe, Henrik	Haldor Topsøe
Van den Bussche, Kurt	UOP

Name	Affiliation
Vogt, Eelco	Albemarle Corporation
Volante, Skip	Merck & Co., Inc.
Volpe, Tony	Symyx
Weider, Paul	Shell
Weinberg, Dr. Henry	Symyx
Wood, Dr. Thomas E.	3M

Appendix E

2005 Committee of Visitors Review Excerpt

The following excerpt is from the 2005 Committee of Visitors review of the Chemical Sciences, Geosciences, and Biosciences Division (CSGB) Catalysis Science Program (formerly known as the "Catalysis and Chemical Transformations" Program).

C. FINDINGS AND RECOMMENDATIONS OF THE CATALYSIS AND CHEMICAL TRANSFORMATIONS SUBPANEL

C.I. EFFICACY AND QUALITY OF THE PROGRAM'S PROCESSES

(a) Solicit, review, recommend, and document proposal actions

Findings:
(1) Reviewers are critical to unbiased and accurate evaluation of proposals.
(2) The review process and associated documentation for actions were usually thorough and appropriate. However, there is room for improvement in review process.
(3) To his credit, the Program Manager is using an informal database.
(4) We were surprised by the paucity of reviewers from industry, given the obvious relevance of this topic.
(5) The DOE is moving toward center or multi-PI interdisciplinary programs. It was not always apparent how the work in individual PI funded programs was distinct from that in multi-PI programs.

Comment:
(1) Continuity in Program Managers is essential for effective program. This has been particular problem with CCT until about three years ago.
(2) The subpanel applauds the Program Manager for significant improvement in creating a coherent and vital program.

Recommendations:
(1) Mandate a request for a list of collaborators and others with a possible conflict of interest as part of grant submission to assist in the selection of reviewers.
(2) Mentors should be listed and not solicited for reviews.
(3) Routinely request suggested and excluded reviewers from the PI.
(4) We strongly recommend creation of a standardized database for reviewers, including: who proposals were sent to, who responded, reasons for not responding (conflict of interest, unresponsive), areas of expertise, evaluation of objectivity and quality of review, timeliness. The COV particularly calls out the inclusion of reviewers from industry.
(5) Mechanisms should be developed to assure a diverse set of reviewers. The Program Manager should further develop a database that includes **diversity**. Use of reviewers from industry in *catalysis* is highly desirable and should be more widely implemented.
(6) Develop a plan for continuity in program management so there are not single-point failure modes for vital programs (*e.g.*, sudden departure of a Program Manager).
(7) As the DOE moves toward center or multi-PI support, it is important to require a section in the proposals on how any new or initiative-driven research relates to other funded research to avoid "double funding".
(8) Consider awarding renewals for longer than three years for exceptional projects, in parallel with the policy for reduced-term renewals in less-compelling cases.
(9) We recommend documenting telephone and verbal communications between PI's, Program Managers, and national lab managers, especially when problems are identified. Further, there should be documentation of follow-up actions.
(10) The subpanel sees no downside to providing verbatim reviewers' comments to individual national lab PI's.
(11) A more formal mechanism for putting national lab scientists on notice for termination is needed.

(b) Monitor active projects and programs

Findings:
(1) The institution of regular contractors meetings has had a positive impact on the overall program. This appears to be one of the main mechanisms for the Program Manager to monitor active progress.
(2) While there is currently a well-defined format for annually reporting on progress, it was not clear how the Program Manager was using this tool (*e.g.*, there was some indication that not all PIs were fully compliant).

APPENDIX E *115*

Recommendations:
(1) Consider holding the contractors meetings at other national meetings to conserve travel expenses?
(2) Examine the efficacy of the annual reporting process.

C.II. EFFECT OF THE AWARD PROCESS ON PORTFOLIOS

(a) the breadth and depth of portfolio elements

Findings:
(1) The *CCT* program supports outstanding science.
(2) The subpanel applauds the Program Manager's efforts to evolve the portfolio elements by responding to (a) community opinions of emerging areas, (b) recommendations resulting from Contractors Meetings, summaries from Council on Chemical Sciences and BESAC workshops, (c) proposal pressure, (d) reading of literature and attending scientific meetings.
(3) The portfolio has evolved towards addressing some of the most challenging aspects of catalysis science. Of particular note is the recent Catalysis Science Initiative.
(4) Some improvements are needed to better inform and focus Program Managers on emerging new areas, needs, and opportunities.
(5) It was apparent to the subpanel that the Program Managers do not have sufficient funds for travel to even a few national conferences. More active participation at scientific meetings would be desirable to accelerate evolution of the portfolios, as well as improve proposal referee base.
(6) The subpanel is concerned with the lack of transparency as regards the administration of funding for PI's with joint national lab-university appointments. The Program Manager does not appear to be able to use all the program management tools employed for other programs. There appears to be confusion and an accompanying skepticism among the general community as regards the size of individual grants, especially for scientists with university appointments. We found it difficult to ascertain the funds provided to individual PI's from the material provided to us.

Comment:
(1) Recommendations from attendees at Contractors Meetings could be self-serving to those currently funded. CCS and BESAC workshops are initiated largely by those other than PM's.

Recommendations:
(1) Workshops should be an efficient and effective means to evolve the Program Manager's portfolio. For example, the Program Managers should have access to suggesting and organizing informal, focused workshops (*e.g.*, Council on

Chemical Sciences). More extensive involvement of non-DOE funded (and non-US) participants would infuse new perspectives and allow a less conflicted set of recommendations.

(2) Mechanisms should be put in place for Program Managers to attend scientific meetings, together with more travel funds, to make the Program Managers more visible and involved in science that they manage, as well as expose them to new thrusts.

(3) A more transparent reporting should be provided the COV for each PI's funding for DOE national lab-university PI's, consistent with that for university PI's not affiliated with a DOE national lab.

(b) the national and international standing of the portfolio elements

Finding:
(1) The DOE *CCT* program is the nation's leading program in catalysis, well represented with national and international awards, ACS awards, National Academy of Sciences memberships, *etc.*

Comments:
(1) For technologically driven research aimed at bringing science to the marketplace, the desirability of multi-investigator, multi-disciplinary funding is well recognized. However, the field of chemical sciences still finds a unique place for the single investigator grant. Through the commitment of time, unfettered by negotiation and administration of a collaborative effort, can a chemist devote the single-minded concentration necessary to perceive, plan, pursue, and solve a problem of singular significance.

(2) Regarding the level of funding of single investigators, funding at the level currently offered through the DOE BES program for single investigator grants poses a significant risk to the maintenance of the excellence that the program has enjoyed. Ideally, funding should allow for the support of at least two, and preferably three persons (post-doctoral or graduate students) per year if the program is going to attract and retain the best PI's. Failure to maintain this level of support will lead to a natural attrition of the very best PI's as they could seek more substantial funding elsewhere. This could lead ultimately to a lessening of the impact of the science accomplished within BES.

Appendix F

Catalysis Science Program Principal Investigators

TABLE F-1 Catalysis Science Program Principal Investigators According to Area of Portfolio, FY 1999–FY 2007

Name	Institution	Area
Abu-Omar, Mahdi	Purdue University	Homogeneous
Adams, Richard	University of South Carolina	HFI
Altman, Eric	Yale University	Surface Science
Angelici, Robert	Ames Laboratory	Homogeneous
Arnold, Frances	California Inst. of Technology	Biorelated
Arnold, John	Lawrence Berkeley National Laboratory	Homogeneous
Atwood, Jim D.	University at Buffalo, State University of New York	Homogeneous
Autrey, S. Thomas	Pacific Northwest National Laboratory	HFI
Bakac, Andreja	Ames Laboratory	Homogeneous
Baker, R. Terry K.	Northeastern University	Heterogeneous
Balbuena, Perla	Texas A&M University	Theory
Barnes, Craig	University of Tennessee	Nanoscience
Barteau, Mark	University of Delaware	Surface Science
Bartels, Ludwig	University of California, Riverside	Catalysis Science Initiative
Bayachou, Mekki	Cleveland State University	Biorelated
Bazan, Guillermo	University of California at Santa Barbara	Homogeneous

Name	Institution	Area
Bell, Alexis T.	University of California, Berkeley	Heterogeneous
Bercaw, John	California Institute of Technology	Homogeneous
Bergman, Robert	University of California, Berkeley	Homogeneous
Brinker, Jeffrey	Sandia National Laboratory	Nanoscience
Brown, Gilbert	Oak Ridge National Laboratory	HFI
Buchanan, III, Archibald	Oak Ridge National Laboratory	Homogeneous
Bullock, Morris	Pacific Northwest National Laboratory	HFI
Campbell, Charles T.	University of Washington	Nanoscience
Carrado, Kathleen	Argonne National Laboratory	Heterogeneous
Casey, Charles	University of Wisconsin, Madison	Homogeneous
Caulton, Kenneth	Indiana University	Homogeneous
Ceyer, Sylvia T.	Massachusetts Institute of Technology	Surface Science
Chen, Jingguang	University of Delaware	Surface Science
Chirik, Paul	Cornell University	Homogeneous
Chisholm, Malcolm	Ohio State University	Homogeneous
Coates, Geoffrey	Cornell University	Homogeneous
Conner, William	University of Massachusetts	Heterogeneous
Cox, David	Virginia Polytechnic Institute	Surface Science
Crabtree, Robert	Yale University	Homogeneous
Crooks, Richard	University of Texas, Austin	Catalysis Science Initiative
Cundari, Tom	University of Northern Texas	Theory
Datye, Abhaya	University of New Mexico	HFI
Davis, Mark	California Inst. of Technology	Catalysis Science Initiative
Davis, Robert	University of Virginia	Heterogeneous
Deem, Michael	Rice University	Theory
Delgass, Nicholas	Purdue University	Catalysis Science Initiative

APPENDIX F

Name	Institution	Area
Diebold, Ulrike	Tulane University	Surface Science
Dubois, Daniel	Pacific Northwest National Laboratory	Homogeneous
Dumesic, James	University of Wisconsin	Heterogeneous
Erlebacher, Johans	Johns Hopkins University	HFI
Evans, William	University of California, Irvine	Homogeneous
Finke, Richard	Colorado State University	Nanoscience
Flytzani-Stephanopoulos, Maria	Tufts University	HFI
Foley, Henry C.	University of Delaware	Other
Ford, Peter C.	University of California, Santa Barbara	Homogeneous
Franz, James	Pacific Northwest National Laboratory	Homogeneous
Friend, Cynthia	Harvard University	Surface Science
Gagne, Michael	University of North Carolina	Homogeneous
Gajewski, J.J.	Indiana University	Other
Gates, Bruce	University of California, Davis	Heterogeneous
Goldman, Alan	Rutgers University	Homogeneous
Goodman, D. Wayne	Texas A&M University	Surface Science
Gorte, Raymond	University of Pennsylvania	Heterogeneous
Grant, David M.	University of Utah	Heterogeneous
Greenbaum, Eias	Oak Ridge National Laboratory	Biorelated
Grey, Clare	Univ. of New York at Stony Brook	Heterogeneous
Guliants, Vadim	University of Cincinnati	Heterogeneous
Gunnoe, Brent	North Carolina State University	Homogeneous
Haller, Gary	Yale University	Heterogeneous
Hartwig, John	University of Illinois, Urbana-Champaign	Homogeneous
Haw, James	University of Southern California	Heterogeneous
Heinz, Tony	Columbia University	Catalysis Science Initiative

Name	Institution	Area
Henderson, Michael	Pacific Northwest National Laboratory	Surface Science
Hess, A.C.	Pacific Northwest National Laboratory	Theory
Hrbek, Jan	Brookhaven National Laboratory	Surface Science
Iglesia, Enrique	University of California, Berkeley	Catalysis Science Initiative
Johnson, Duane	University of Illinois, Urbana-Champaign	Catalysis Science Initiative
Jones, Christopher	Georgia Tech University	Catalysis Science Initiative
Jones, William	University of Rochester	Homogeneous
Jordan, Richard	University of Chicago	Homogeneous
Katz, Alexander	University of California, Berkeley	Heterogeneous
Kemp, Richard	University of New Mexico	Homogeneous
Kern, Dorothee	Brandeis University	Biorelated
Kitchin, John	Carnegie Mellon University	Theory
Klier, Kamil	Lehigh University	Heterogeneous
Koel, Bruce E.	University of Southern California	Surface Science
Kubas, Gregory	Los Alamos National Lab	Homogeneous
Kung, Harold	Northwestern University	Heterogeneous
Landis, Clark	University of Wisconsin	Homogeneous
Linic, Suljo	University of Michigan	Theory
Lobo, Raul	University of Delaware	Heterogeneous
Long, Jeffrey	University of California, Berkeley	Nanoscience
Louie, Janis	University of Utah	Homogeneous
Maatta, Eric A.	Kansas State University	Homogeneous
Madey, Theodore	Rutgers University	Surface Science
Madix, Robert J.	Stanford University	Surface Science
Malinakova, Helena	University of Kansas	Homogeneous
Marks, Tobin	Northwestern University	Homogeneous

APPENDIX F

Name	Institution	Area
Maverick, Andrew	Louisiana State University	Nanoscience
Mavrikakis, Manos	University of Wisconsin	Theory
McCabe, Clare	Vanderbilt University	Theory
McFarland, Eric	University of California, Santa Barbara	Heterogeneous
Mindiola, Daniel	Indiana University	Homogeneous
Musaev, Djamaladdin	Emory University	Catalysis Science Initiative
Neurock, Matthew	University of Virginia	Theory
Norton, Jack	Columbia University	Homogeneous
Nuckolls, Colin	Columbia University	Nanoscience
Nuzzo, Ralph	University of Illinois, Urbana-Champaign	Catalysis Science Initiative
Overbury, Steven	Oak Ridge National Laboratory	Nanoscience
Oyama, S. Ted	Virginia Polytechnic Institute	Heterogeneous
Ozerov, Oleg	Brandeis University	Homogeneous
Ozkan, Umit	Ohio State University	HFI
Painter, Paul	Pennsylvania State University	
Parkin, Gerard	Columbia University	Homogeneous
Peden, Charles	Pacific Northwest National Laboratory	Catalysis Science Initiative
Pendse, Hemant	University of Maine	Biorelated
Pfefferle, Lisa	Yale University	HFI
Pruski, Marek	Ames Laboratory	Catalysis Science Initiative
Que, Lawrence	University of Minnesota	Biorelated
Rabideau, P.W.	Louisiana State University	Nanoscience
Rahman, Talat	University of Central Florida	Theory
Rappe, Andrew	University of Pennsylvania	Theory
Rathke, Jerome	Argonne National Laboratory	Homogeneous
Rauchfuss, Thomas	University of Illinois	Homogeneous
Raymond, Kenneth	University of California, Berkeley	Homogeneous

Name	Institution	Area
Rehr, John	University of Washington	Theory
Resasco, Daniel	University of Oklahoma	Nanoscience
Ribeiro, Fabio	Purdue University	Surface Science
Roddick, Dean	University of Wyoming	Homogeneous
Rodriguez, Jose	Brookhaven National Laboratory	HFI
Rothwell, Ian P.	Purdue Research Foundation	Homogeneous
Rybak-Akimova, Elena	Tufts University	Homogeneous
Sachtler, Wolfgang	Northwestern University	Heterogeneous
Schmidt, Lanny	University of Minnesota	Heterogeneous
Schneider, William	University of Notre Dame	Theory
Schrock, Richard	Massachusetts Institute of Technology	Homogeneous
Schwarz, Udo	Yale University	Surface Science
Scott, Lawrence	Boston College	Homogeneous
Scott, Susannah	University of California, Santa Barbara	Catalysis Science Initiative
Sen, Ayusman	Pennsylvania State University	Homogeneous
Seshadri, Ram	University of California, Santa Barbara	HFI
Shao-Horn, Yang	Massachusetts Institute of Technology	HFI
Sharp, Paul	University of Missouri	Homogeneous
Shelnutt, John	University of Georgia	Nanoscience
Sholl, David	Carnegie Mellon University	Heterogeneous
Sigman, M.E.	Oak Ridge National Laboratory	Homogeneous
Sneddon, Larry G.	University of Pennsylvania	Nanoscience
Somorjai, Gabor	Lawrence Berkeley National Laboratory	Nanoscience
Stahl, Shannon	University of Wisconsin, Madison	Homogeneous/Biorelated
Stair, Peter	Northwestern University and Argonne National Lab	Surface Science
Suib, Stephen	University of Connecticut	Heterogeneous

Name	Institution	Area
Sutter, Peter	Brookhaven National Laboratory	Surface Science
Sygula, Andrzej	Mississippi State University	Nanoscience
Theopold, Klaus	University of Delaware	Homogeneous
Thumauer/Wagner	Argonne National Laboratory	Catalysis Science Initiative
Tong, YuYe	Georgetown University	HFI
Trenary, Michael	University of Illinois, Urbana-Champaign	HFI
Tysoe, Wilfred T.	University of Wisconsin	Surface Science
Vicic, David	University of Hawaii	Heterogeneous
Vlachos, Dionisios	University of Delaware	Theory
Vohs, John	University of Pennsylvania	Surface Science
Vollhardt, K.P.C	Lawrence Berkeley National Laboratory	Homogeneous
Wachs, Israel	Lehigh University	Heterogeneous
Wang	Washington State University	Catalysis Science Initiative
Wayland, Bradford	University of Pennsylvania	Homogeneous
Weaver, Jason	University of Florida	Surface Science
Weinberg	University of California, Santa Barbara	Surface Science
White, John M.	University of Texas, Austin	Surface Science
White, Michael G.	Brookhaven National Laboratory	Nanoscience
Winans, Randall	Argonne National Laboratory	Homogeneous
Wu, Ruqian	University of California, Irvine	Theory
Yates, John	University of Virginia	Surface Science
Zaera, Francisco	University of California, Riverside	Catalysis Science Initiative

SOURCE: Basic Energy Sciences Catalysis Science Program staff.

TABLE F-2 Catalysis Science Program Principal Investigators, FY 2008 (investigators new in 2008 shown in italics)

Name	Institution
Abu-Omar, Mahdi	Purdue University
Adzic, Radoslav	*Brookhaven National Laboratory*
Altman, Eric	Yale University
Arnold, Frances	California Inst. of Technology
Auerbach, Scott	*University of Massachusetts*
Autrey, S. Thomas	Pacific Northwest National Laboratory
Bakac, Andreja	Ames Laboratory
Balbuena, Perla	Texas A&M University
Barnes, Craig	University of Tennessee
Barteau, Mark	University of Delaware
Bartels, Ludwig	University of California, Riverside
Bazan, Guillermo	University of California, Santa Barbara
Bell, Alexis T.	University of California, Berkeley
Bercaw, John	California Institute of Technology
Bergman, Robert	University of California, Berkeley
Brinker, Jeffrey	Sandia National Laboratory
Britt, Philip	*Oak Ridge National Laboratory*
Brown, Gilbert	Oak Ridge National Laboratory
Buchanan, III, Archibald	Oak Ridge National Laboratory
Bullock, Morris	Pacific Northwest National Laboratory
Camillone, III, Nicholas	*Brookhaven National Laboratory*
Campbell, Charles T.	University of Washington
Casey, Charles	University of Wisconsin, Madison
Chan, Siu-Wai	*Columbia University*
Chang, Christopher	*University of California, Berkeley*
Chen, Jingguang	University of Delaware
Chirik, Paul	Cornell University
Chisholm, Malcolm	Ohio State University
Chmelka, Bradley	*University of California, Santa Barbara*
Coates, Geoffrey	Cornell University
Conner, William	University of Massachusetts
Cox, David	Virginia Polytechnic Institute
Crabtree, Robert	Yale University
Crooks, Richard	University of Texas at Austin
Cundari, Tom	University of Northern Texas
Dai, Sheng	*Oak Ridge National Laboratory*
Datye, Abhaya	University of New Mexico
Davis, Mark	California Inst. of Technology
Davis, Robert	University of Virginia
Deem, Michael	Rice University

APPENDIX F

Name	Institution
Delgass, Nicholas	Purdue University
Diebold, Ulrike	Tulane University
Dixon, David	University of Alabama
Dubois, Daniel	Pacific Northwest National Laboratory
Dumesic, James	University of Wisconsin
Eckert, Juergen	Los Alamos National Laboratory
Erlebacher, Johans	Johns Hopkins University
Evans, William	University of California, Irvine
Finke, Richard	Colorado State University
Flytzani-Stephanopoulos, Maria	Tufts University
Franz, James	Pacific Northwest National Laboratory
Frei, Heinz	Lawrence Berkeley National Laboratory
Frenkel, Anatoly	Yeshiva University
Friend, Cynthia	Harvard University
Gagne, Michael	University of North Carolina
Gates, Bruce	University of California, Davis
Gellman, Andrew	Carnegie Mellon University
Goldberg, Karen	University of Washington
Goldman, Alan	Rutgers University
Goodman, D. Wayne	Texas A&M University
Gorte, Raymond	University of Pennsylvania
Grey, Clare	University of New York at Stony Brook
Guliants, Vadim	University of Cincinnati
Gunnoe, Brent	North Carolina State University
Hagaman, Edward	Oak Ridge National Laboratory
Haller, Gary	Yale University
Hartwig, John	University of Illinois, Urbana-Champaign
Haw, James	University of Southern California
Heinz, Tony	Columbia University
Henderson, Michael	Pacific Northwest National Laboratory
Hill, Craig	Emory University
Hrbek, Jan	Brookhaven National Laboratory
Hupp, Joseph	Northwestern University
Iglesia, Enrique	University of California, Berkeley
Johnson, Duane	University of Illinois, Urbana-Champaign
Jones, Christopher	Georgia Tech University
Jones, William	University of Rochester
Jordan, Richard	University of Chicago
Katz, Alexander	University of California, Berkeley
Kemp, Richard	University of New Mexico
Kern, Dorothee	Brandeis University
Kitchin, John	Carnegie Mellon University

Name	Institution
Klingler, Robert	*Argonne National Laboratory*
Kubas, Gregory	Los Alamos National Lab
Kung, Harold	Northwestern University
Landis, Clark	University of Wisconsin
Lauterbach, Jochen	*University of Delaware*
Lin, Victor	*Iowa State University*
Linehan, John	Pacific Northwest National Laboratory
Linic, Suljo	University of Michigan
Liu, Jun	*Pacific Northwest National Laboratory*
Liu, Meilin	*Georgia Tech University*
Lobo, Raul	University of Delaware
Long, Jeffrey	University of California, Berkeley
Louie, Janis	University of Utah
Madey, Theodore	Rutgers University
Malinakova, Helena	University of Kansas
Marks, Tobin	Northwestern University
Marshall, Christopher	*Argonne National Laboratory*
Maverick, Andrew	Louisiana State University
Mavrikakis, Manos	University of Wisconsin
McFarland, Eric	University of California, Santa Barbara
Mindiola, Daniel	Indiana University
Mullins, David	*Oak Ridge National Laboratory*
Musaev, Djamaladdin	Emory University
Neurock, Matthew	University of Virginia
Nguyen, SonBinh	*Northwestern University*
Norton, Jack	Columbia University
Nuckolls, Colin	Columbia University
Nuzzo, Ralph	University of Illinois, Urbana-Champaign
O'Brien, Stephen	*Columbia University*
Overbury, Steven H.	Oak Ridge National Laboratory
Oyama, S. Ted	Virginia Polytechnic Institute
Ozerov, Oleg	Brandeis University
Ozkan, Umit	Ohio State University
Parkin, Gerard	Columbia University
Peden, Charles	Pacific Northwest National Laboratory
Pfefferle, Lisa	Yale University
Pruski, Marek	Ames Laboratory
Que, Lawrence	University of Minnesota
Rahman, Talat	University of Central Florida
Rappe, Andrew	University of Pennsylvania
Rathke, Jerome	Argonne National Laboratory
Rauchfuss, Thomas	University of Illinois
Raymond, Kenneth	University of California, Berkeley

Name	Institution
Rehr, John	University of Washington
Resasco, Daniel	University of Oklahoma
Ribeiro, Fabio	Purdue University
Rodriguez, Jose	Brookhaven National Laboratory
Rybak-Akimova, Elena	Tufts University
Sadow, Aaron	*Iowa State University*
Schmidt, Lanny	University of Minnesota
Schneider, William	University of Notre Dame
Schrock, Richard	Massachusetts Institute of Technology
Schwarz, Udo	Yale University
Scott, Lawrence	Boston College
Scott, Susannah	University of California, Santa Barbara
Selloni, Annabella	*Princeton University*
Sen, Ayusman	Pennsylvania State University
Seshadri, Ram	University of California, Santa Barbara
Shao-Horn, Yang	Massachusetts Institute of Technology
Sharp, Paul	University of Missouri
Sherrill, David	*Georgia Tech*
Sholl, David	Carnegie Mellon University
Somorjai, Gabor	Lawrence Berkeley National Laboratory
Stahl, Shannon	University of Wisconsin, Madison
Stair, Peter	Northwestern University and Argonne National Laboratory
Suib, Stephen	University of Connecticut
Sutter, Peter	Brookhaven National Laboratory
Sygula, Andrzej	Mississippi State University
Theopold, Klaus	University of Delaware
Tilley, T. Don	*University of California, Berkeley*
Tong, YuYe	Georgetown University
Trenary, Michael	University of Illinois, Urbana-Champaign
Tysoe, Wilfred T.	University of Wisconsin
Vicic, David	University of Hawaii
Vlachos, Dionisios	University of Delaware
Vohs, John	University of Pennsylvania
Wachs, Israel	Lehigh University
Wang, Yong	*Pacific Northwest National Laboratory*
Wayland, Bradford	University of Pennsylvania
Weaver, Jason	University of Florida
Weck, Marcus	*New York University*
White, Michael G.	Brookhaven National Laboratory
Woo, Keith	*Iowa State University*
Wu, Ruqian	University of California, Irvine
Yang, Judith	*University of Pittsburgh*

Name	Institution
Yates, John	University of Virginia
Zaera, Francisco	University of California, Riverside

SOURCE: Basic Energy Sciences Catalysis Science Program staff.